Lecture Notes in Bioinformatics 6575

Subseries of Lecture Notes in Computer Science

W0192965

Corrado Priami   Ralph-Johan Back
Ion Petre   Erik de Vink (Eds.)

Transactions on Computational Systems Biology XIII

 Springer

Series Editors

Sorin Istrail, Brown University, Providence, RI, USA
Pavel Pevzner, University of California, San Diego, CA, USA
Michael Waterman, University of Southern California, Los Angeles, CA, USA

Editor-in-Chief

Corrado Priami
The Microsoft Research - University of Trento
Centre for Computational and Systems Biology
Piazza Manci, 17, 38050 Povo (TN), Italy
E-mail: priami@dit.unitn.it

Guest Editors

Ralph-Johan Back
Ion Petre
Åbo Akademi University
Department of Information Technologies
Joukohaisenkatu 3-5, 20520 Turku, Finland
E-mail: {backrj,ipetre}@abo.fi

Erik de Vink
Technische Universiteit Eindhoven
Den Dolech 2, Eindhoven, The Netherlands
E-mail: evink@win.tue.nl

ISSN 0302-9743 (LNCS)   e-ISSN 1611-3349 (LNCS)
ISSN 1861-2075 (TCSB)   e-ISSN 1861-2083 (TCSB)
ISBN 978-3-642-19747-5   ISBN 978-3-642-19748-2 (eBook)
DOI 10.1007/978-3-642-19748-2
Springer Heidelberg Dordrecht London New York

Library of Congress Control Number: 2011922750

CR Subject Classification (1998): J.3, F.1-2, F.4, I.6, I.2, C.1.3

Typesetting: Camera-ready by author, data conversion by Scientific Publishing Services, Chennai, India

Printed on acid-free paper

Springer is part of Springer Science+Business Media (www.springer.com)

Preface

The many facets of life are reflected by the multitude of dimensions of systems biology research at present. Current modeling and analysis approaches to a systematic understanding of biological phenomena range from quantitative to qualitative, from discrete to continuous, from deterministic to stochastic, from concrete detailed biological case studies to abstract bio-inspired computing paradigms. This special issue of the Transactions on Computational Systems Biology on *Computational Models for Cell Processes* also mirrors the rich variety of the field.

The volume is based on the CompMod workshop that took place in Eindhoven, the Netherlands, on November 2, 2009. Previously held in Turku, Finland, the workshop was organized for the second time, now as a satellite event of the 16th International Symposium on Formal Methods, part of FMweek, running from November 2 to 6, 2009 in Eindhoven. The CompMod workshop aims to foster a platform gathering researchers in formal methods and related fields interested in the wealth of challenges and opportunities in systems biology. A specific interest is expressed for papers discussing biological processes requiring special tools and techniques not investigated so far in the context of formal methods, as well as extensions of formal methods formalisms introduced to improve their applicability to biology. For this special issue there has been an additional open call for paper submissions, with a separate peer-reviewing process.

The papers included illustrate the broad span of aspects of modeling and analysis of biological systems: evolution of a cell population with selection based on toxin resistance; a quantitative and tool-supported interpretation of flow abstraction in the Systems Biology Graphical Notation; an analytic approach to dynamic simulation of deformable biological structures; a new stochastic simulation algorithm reconsidering the delay-as-duration principle; a process algebraic case study on ammonium transport in plant-fungus symbiosis; iterative variable elimination for steady state equations using algebraic modules in the analysis of metabolic networks. From different points of view and following various approaches the papers cover a wide range of topics in Systems Biology, addressing the dynamics we begin to unravel and computational principles that we start to identify.

This issue also includes two regular papers by Wallace and Wallace on the heritability of complex diseases and by Paulevé et al. on the dynamics of gene regulatory networks.

December 2010

Ralph-Johan Back
Ion Petre
Corrado Priami
Erik de Vink

LNCS Transactions on
Computational Systems Biology –
Editorial Board

Table of Contents

Evolutionary Dynamics of a Population of Cells with a Toxin Suppressor Gene

Antti Häkkinen[1], Fred G. Biddle[2], Olli-Pekka Smolander[1],
Olli Yli-Harja[1], and Andre S. Ribeiro[1,3]

[1] Computational Systems Biology Research Group,
Tampere University of Technology, Finland
[2] Department of Medical Genetics, Institute of Maternal and Child Health,
Faculty of Medicine, University of Calgary, Canada
[3] Center for Computational Physics, University of Coimbra,
P-3004-516 Coimbra, Portugal

Abstract. Environmental changes are known to trigger evolutionary changes, e.g. by favoring higher mutation rates. We study the evolutionary dynamics of a delayed stochastic genetic circuit using a simulator developed for this aim. We model a cell population subject to selection and environmental changes. Each cell contains a self-repressing gene whose protein degrades a toxin. Allowing mutations, we study the adaptability of this circuit and how the genotypic and phenotypic diversities of the population evolve. Neutral mutations and equally beneficial evolutionary pathways are found to generate complex phenotypic distributions. We find optimal mutation rates dependent on the amount of toxin and show that shifting environmental conditions trigger transient increases in diversity. The results support the hypothesis that evolvability is a selectable trait.

1 Introduction

Organisms adapt to a wide range of unpredictable environmental changes. Genotypic and phenotypic diversity, which play a major role in the organisms' potential to adapt to changes, are likely to be heritable and to be partially responsible for organisms' robustness [1]. Especially in prokaryotes, noise in gene expression is a key source of phenotypic diversity [2,3]. Another source is the interaction between organisms and the environment [4].

In unstable environments, organisms are likely to need higher mutation rates, unlike in more stable conditions, as high mutation rates tend to cause the accumulation of deleterious mutations [5]. Selection can only act when there is variability within a population [6]. Since variability depends on the mutation rate, the ability to control this rate is a selectable trait. In support of this hypothesis, bacterial mutation rates were found to increase in the initial stages of colonization of a mouse gut [7], decreasing afterwards.

The ability to generate heritable phenotypic variation is a selectable trait [1]. One such case has been characterized in *Bacillus subtilis*, which has probabilistic and transient cellular differentiation, dependent on the environment [8]. The

C. Priami et al. (Eds.): Trans. on Comput. Syst. Biol. XIII, LNBI 6575, pp. 1–12, 2011.
© Springer-Verlag Berlin Heidelberg 2011

probability of being in either state is stationary within a given external condition, and is determined by the noise in ComK expression level [8]. Reduction of the noise decreases the number of competent cells, suggesting that noise-driven genetic mechanisms can evolve [9].

Here, we study the evolutionary dynamics of a self-repressing gene responsible for coping with a toxin, and whose dynamics is driven by a delayed stochastic simulation algorithm, at the single molecule level. Each model cell has a gene responsible for resistance to tetracycline that has been characterized in bioluminescent *Escherichia coli* K-12 [10]. Tetracycline resistance is regulated by the tetA promoter and the TetR protein, which acts as a self-repressor. In the absence of tetracycline, the TetR protein binds to the promoter and represses own expression that was induced when tetracycline was added [10]. The model environment consists of the amount of exposure of each cell to the toxin tetracycline.

We address the following questions: Do genotypic and phenotypic diversity depend on environmental conditions? How does the rate of change of the environment affect these diversities? Are there optimal mutation rates for a given environment?

2   Methods

We simulate, at the single cell level, cell populations that are subject to selection at the end of each generation. The dynamics of each cell is driven by the delayed Stochastic Simulation Algorithm [11], based on the original SSA [12], and implemented in SGNSim [13]. The model of gene expression [14] accounts for stochastic fluctuations and, by using multiple-time delayed reactions, it accounts for the fact that transcription and translation are multiple-step processes and take non-negligible time to be completed once they are initiated. The model was validated by matching measurements of the time series of gene expression at the single molecule level [15,16]. Time delayed reactions are represented as: $A + C \rightarrow A(\tau_1) + B(\tau_2) + D$. When the reaction occurs, C is instantaneously consumed, and D is instantaneously produced. Substance A is not consumed but it is placed on a waitlist until it is released after τ_1 s, while a new substance B is produced τ_2 s after the reaction occurs [11,13].

To model mutations and cell selection, we developed and implemented a wrapper program for SGNSim, named "CellSelector". CellSelector allows running multiple independent simulations of single gene models in parallel for a specified time length, which corresponds to the cells lifetime. In our simulations, for each set of conditions, we run 100 independent threads. The simulation of the dynamics of each cell is seeded with a unique seed to initialize the random number generator, responsible for the generation of the stochasticity of the simulation according to the SSA, thus guaranteing that the cells in each generation have unique trajectories in the state space. The simulator program is available upon request.

After the fixed lifetime of the cells of a generation is past, the final state of the each cell is observed. Selection then occurs, based on these states. Namely, the cells are sorted by a fitness measure *fit*, and those belonging to the least fit q-quantile are eliminated, while the others are used to produce two or more duplicates for the subsequent cell generation. In our simulations we always eliminate 50% of the cells at the end of each generation, and make two duplicate cells out of each of the remaining cells that will constitute the cell population of the next generation.

The initial state of the daughter cells is set to be identical to the final state of their mother cell (with a new random seed being generated for each daughter cell). This implies that any mutations accumulated by the mother cell are present in the daughter cells. Only the fitness measure is set to zero at the beginning of each cell lifetime.

When toxin is present, it binds to protein p (even when the protein is bound to the promoter), which therefore can no longer repress the gene, allowing transcription to take place. Being a stochastic system, the higher the number of toxins in the cell, the more likely it is that the promoter is free to transcribe. At any given moment, we define "environmental conditions" as the number of toxins that the cell is subject to.

The environmental conditions determine both how much time cells are subject to the toxin and the amount of toxin. Toxin ("X") is introduced in the cells via reaction (1) at rate c_{pois} (the value of this rate defines the environmental condition at any given moment) and degrades via reaction (2) at rate d_{pois}.

These reactions are only active when the cell is subject to toxin, and they impose approximately constant amount of toxin over time during these periods. Tuning c_{pois} allows controlling such amount:

$$\xrightarrow{c_{pois}} X \tag{1}$$

$$X \xrightarrow{d_{pois}} \emptyset \tag{2}$$

A cell's fitness is measured throughout its life. The toxin is assumed to be harmful. Excess of protein is also assumed harmful, since in the case of the gene studied here it leads to cell death due to loss of membrane potential [17]. Thus, we assume that the goal of each cell and the selection process is to simultaneously decrease the amounts of toxin and protein. Finally, in order to inactive a toxin X, a protein p needs to bind to it, forming the complex Xp. The number of Xp complexes is a good indicator of the fitness of the cell.

Combining these conditions, fitness is stochastically measured by reaction (3) (the symbol * indicates that the reactant is not consumed in the reaction, although it affects the propensity of the reaction [13]):

$$^*X +^* p +^* Xp \xrightarrow{c_{fit}} fit \tag{3}$$

The propensity ($Prop(4)$) [12] of reaction (3) is calculated at each step of the stochastic simulation by equation (4):

$$Prop(4) = c_{fit}r \times ([X]+1)^{-1} \times ([p]+1)^{-1} \times ([Xp]+1) \tag{4}$$

Note, from reaction (3) and the formula used to compute its propensity (4), that the more toxins X and proteins p exist in the cell, the less fitness units, fit, will be produced. For that to be possible in the simulation, following the protocols of SGNSim [13], we introduced in the left hand side of the reaction X and p, so as to allow the propensity of the reaction to be inversely dependent on the amounts of these two substances, since the speed of production of fit is determined by the propensity (defined in (4)) [12].

Reaction (3) doesn't affect the cells' dynamics since no substance is consumed and the product is not a substrate to any reaction. Its propensity [12] determines how many fitness units are produced and is computed according to equation (4), in agreement with the fitness conditions proposed. All cells have zero fitness units in the beginning of their life.

To the best of our knowledge, this method of computing the fitness of a genetic circuit at runtime by introducing a stochastic reaction in the system has not been previously used. It is therefore important to note that the dynamics of the other reactions in the system are not affected in any way, and that the value of fitness has a stochastic component. According to the SSA, the number of times and the moments when a reaction occurs is solely determined by its propensity at each moment [12]. In our model, the dynamics of all other reactions (i.e., the number and the moment of occurrence of the reactions) are not affected by the reaction producing fitness units because it does not consume or produce any substances associated to the other reactions, thereby not affecting their propensities at any moment.

Further note that this is also true for reactions 13 and 14, since, as seen later, they do not consume any substances affecting the propensities of the other reactions in the system. In practical terms, our method is equivalent to, e.g., calculating fitness at runtime by having two parallel simulations ongoing simultaneously (one for the system, another for fitness calculation) with the latter being informed of the state of the first at each step.

Additionally, it is noted that while the expression of the propensity of reaction 3 differs from common expressions of propensity of regular chemical reactions (i.e. linear dependence on each substrate), this does not affect the dynamics or functioning of the SSA or the simulator. SGNSim [13] uses the formula to obtain a real value of propensity at each moment which, as in the other reactions, determines the reaction's stochastic rate of occurrence.

Gene expression is modeled by multiple time-delayed reactions, one for transcription (5) by RNA polymerase ($RNAp$), with a stochastic rate constant k_t, and one for translation (6) by ribosomes (rib), with a stochastic rate constant k_{tr}, according to the model proposed in [14]. As mentioned, in these reactions, the delays are represented explicitly. E.g. in reaction 5, the notation "$RBS(2)$" denotes that the ribosome binding site is only produced and introduced in the system $2s$ after the reaction occurs.

Decay reactions degrade p, (7 and 10) and RNA's (represented by their ribosome binding site, RBS [16]) via reaction (8). Reactions (9) model the binding and unbinding of the self-repressor protein to the gene promoter region (Pro).

The binding of X to p when free or when bound to the promoter [10] is modeled by reactions (11) and (12).

$$Pro + RNAp \xrightarrow{k_t} Pro(2) + RNAp(40) + RBS(2) \tag{5}$$

$$RBS + rib \xrightarrow{k_{tr}} RBS(2) + rib(20) + p(50) \tag{6}$$

$$p \xrightarrow{d_p} \emptyset \tag{7}$$

$$RBS \xrightarrow{d_{RBS}} \emptyset \tag{8}$$

$$Pro + p \underset{k_{rep}}{\overset{k_{unrep}}{\rightleftarrows}} Pro.p \tag{9}$$

$$Pro.p \xrightarrow{d_p} Pro \tag{10}$$

$$X + p \xrightarrow{k_{pdes}} X.p \tag{11}$$

$$X + Pro.p \xrightarrow{k_{pdes}} X.p + Pro \tag{12}$$

We assume that the gene is subject to mutations, and that these affect the rate of transcription as well as the strength of repression, since these rates are those that most directly affect the rate of production of the protein. Thus, the rates subject to changes due to mutations in the gene sequence (initiation and elongation regions) are k_t, k_{rep}, and k_{unrep}. Affinity between promoter and protein determines k_{rep} and k_{unrep}, while transcription initiation (k_t) is sequence dependent [18].

We use virtual substances [13] to implement at runtime the effects of mutations in a cell's dynamics. The propensity of reactions (5) and (9) is computed as follows. Let K be the original rate constant, n_{up} be a virtual substance that increases the reaction propensity if its quantity increases, and n_{down} a virtual substance that decreases the propensity if its quantity increases. To do this, the propensity of the reaction, P, is computed by (13):

$$P = K \times (n_{up} + 1) \times \frac{1}{1 + n_{down}} \tag{13}$$

The propensity can be varied at runtime by reactions (14) and (15). A pair of reactions (14) and (15) is added for each rate constant subject to changes due to mutations:

$$\emptyset \xrightarrow{k_{mut}} M \times n_{up} \tag{14}$$

$$\emptyset \xrightarrow{k_{mut}} M \times n_{down} \tag{15}$$

In reactions (14) to (15), tuning k_{mut} allows varying the rate of occurrence of mutations and by tuning M (number of molecules created in one reaction) one can set the extent of the variation in the propensity caused by one mutation. When any of the two reactions, (14) and (15) occur, n_{up} or n_{down} vary, thus changing the propensity of the reaction they affect (either transcription, repression or unrepression). If the change improves the cell fitness, this cell is likely to be selected for duplication at the end of its lifetime.

It is noted that the rates subject to mutation, i.e. k_t, k_{rep}, and k_{unrep}, allow varying both the mean expression as well as the noise strength of the protein level. Also, due to the existence of the delay on the promoter release, identical ratios between k_{rep} and k_{unrep} will produce the same mean expression level, but the RNA and the protein levels will have different noise strengths [19].

3   Results

Reactions (1-12), (14) and (15) are implemented in each cell. The unit of time delays is second (s) and the unit of rate constants is s^{-1}. Since we model a gene from *E. coli* [10] the parameters values are set accordingly, e.g, translation initiation ($k_{tr} = 0.0005$), RNA decay ($d_{RBS} = 0.005$), and protein decay ($d_p = 0.0004$) [16]. The same applies to the time delays in transcription (5) and translation (6). The values of the delays are set according to known kinetic parameters of transcription and translation in *E. coli* (for a detailed justification and derivation of the values of these delays please refer to [16]).

Cell division (and selection) occurs at each 1800 s, which is the average division time of *E. coli*. Additionally, we set k_{pdes} to 0.01 which is within realistic parameter values [20], and c_{fit} to 1.

To impose an average toxin concentration in the cell of, e.g., 10 molecules X, we set $c_{pois} = 0.1$ and $d_{pois} = 0.01$. In a following simulation, c_{pois} will be varied to subject the cells to different environments at runtime.

The rates that are varying due to mutations are initially set to: $k_t = 0.0025$, $k_{rep} = 10^{-4}$, and $k_{unrep} = 0.1$. Real mutation rates in *E. coli* are $\sim 10^{-7}$ per cell division, but vary significantly depending on the conditions to which cells are subject [21]. We vary this rate (k_{mut}) to study its effects. Unless stated otherwise, we model 100 cells per generation for 100 generations (100G).

We first tested the effects of varying M, with $k_{mut} = 10^{-4}$. Cells are subject to toxin for periods of 10G with $c_{pois} = 0.1$ and $d_{pois} = 0.001$, interrupted by periods of 10G not subject to toxin. For $M \leq 2$, mutations effects are below the noise level. Increases in the population's fitness are only due to selection. For $2 < M < 10000$ the average fitness increases for several generations and reach a maximum value, equal for all values of M. It takes from 10G to 50G to reach the maximum fitness.

Next we varied k_{mut} from 10^{-7} to 1. We set $M = 10$, so that mutations cause phenotypic changes significantly above the noise level. For $k_{mut} < 10^{-6}$ and $k_{mut} > 0.1$, fitness only improves due to selection. In the first case, the number of mutations that can occur within 100G is not sufficient to allow the cells to adapt to the change and, in the second case, due to a too high mutation rate, the selection that occurs at each generation is not sufficient to prevent the accumulation of harmful mutations.

Setting k_{mut} to 10^{-4} and M to 10, we now analyze the phenotypic diversity in a simple environment, namely, where no toxin is present in the first 100G and where, from G100 to G200, cells are subject to toxin ($c_{pois} = 0.1$ and $d_{pois} = 10^{-3}$) and, finally, no toxin is present thereafter.

To quantify the diversity of the values of the rate constants subject to changes due to mutations over the population, we compute the ratio between standard deviation and mean value of these rates, i.e. the coefficient of variation, CV, in each generation.

In Fig. 1 we plotted the CV of k_t, k_{rep} and k_{unrep} for 300G. All CV's were comparatively high, i.e. the population had higher phenotypic and genotypic diversities, after environmental changes and in the initial transient, from G1 to $\sim$G30, where cells are adapting to an environment without toxin (evolving towards diminishing the number of proteins by decreasing k_t and/or increasing k_{rep}). The other moments are from G100 to G140 and from G200 to G230, the adaptation periods to the environmental changes (toxin introduced at G100 and removed at G200). Thus, changes in the environment trigger transient increases in genotypic and phenotypic diversities in the populations of model cells.

From these results we conclude that, even assuming fixed mutation rates for simplicity, environmental changes are likely to enhance the degree of variability of a population. When the environmental conditions change, cells that were optimally adapted are no longer as fit as before. Thus, recently mutated cells have greater chances to be fitter in variable environments (in comparison to non mutated ones) than in stable ones.

In the periods that the cells are fittest (from G30 to G100, from G140 to G200, and after G230), the population maintains a considerable diversity. This is due both to the continuous appearance of mutated cells (usually removed in subsequent generations) and neutral mutations, e.g., causing equal increases in k_{rep} and k_{unrep}, which does not change the average time the gene is repressed.

Neutral mutations are one of the causes for the emergence of complex phenotypic distributions (e.g., bimodal) and one example is shown in Fig. 2. In this case some cells have higher k_{rep} than the rest of the population but also higher k_{unrep} (not shown). Another way for genotypic and consequent phenotypic bifurcations to appear is when distinct evolutionary pathways have identical fitness, e.g., instead of increasing k_{rep}, decreasing the transcription rate k_t also diminishes the number of p's in the cell).

Finally, we subject populations of initially identical cells to various environments and measure the average fitness over 100 generations. Toxin is introduced at random moments for random time durations. The transition rate between presence and absence of toxin is set to $1/(10 \times T)$ (where T is the cells lifetime). The environments differ in the amount of toxin present. We set c_{pois} to 0.001, 0.1 and 1, while d_{pois} is kept at 0.01. The results are shown in Fig. 3.

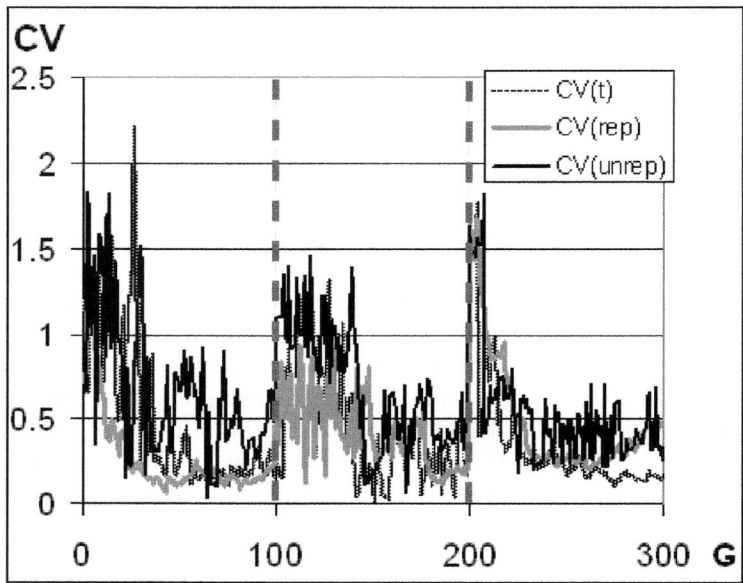

Fig. 1. Coefficient of variation (CV) of rates subject to mutation in 1 simulation of 300 cell generations. Toxin absent from G1 to G100 and from G200 to G300. Vertical dashed lines represent the moments when the environmental conditions changed.

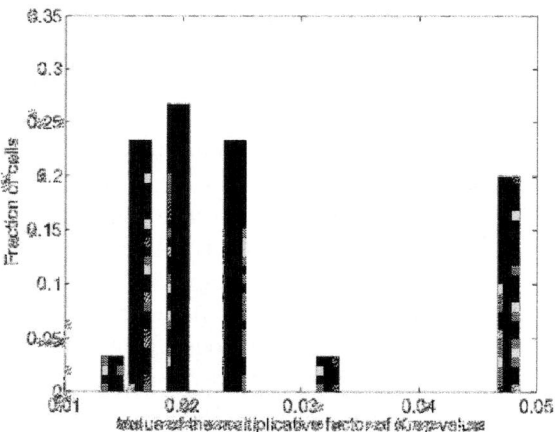

Fig. 2. Population genotypic distribution at G105 of the evolved values of the multiplicative factor of k_{rep}

There are optimal mutation rates (Fig. 3) whose values depend on the environmental conditions, i.e., amounts of toxin. Because the model cells are initially not well adapted to the presence of virtually no toxin, the higher the amount of toxin introduced, the higher must be the mutation rate. That is, a determining

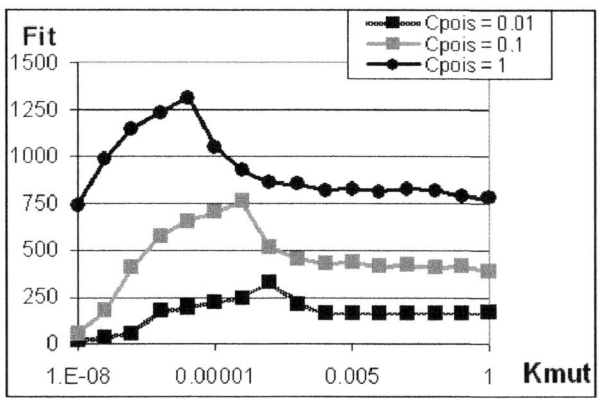

Fig. 3. Average fitness in 100 G for various toxin dosages and mutation rates

factor of the value of optimal mutation rates is the necessary genotypic and phenotypic change to reach maximum fitness. Initially, less well-adapted cells require higher mutation rates to rapidly create diversity from which fitter cells can be selected from. This suggests that it is advantageous for cells to tune or evolve mutation rates depending on the environmental conditions and shifting of these conditions.

Two factors contribute to the existence of optimal mutation rates of this gene: if mutation rates are too low, beneficial mutations do not occur fast enough to improve the population's fitness in reasonable time and, if mutation rates are too high, selection is not sufficiently fast to prevent the accumulation of harmful mutations.

4   Conclusions and Discussion

We implemented in model cells a stochastic model with delays of a self-repressing gene responsible for Tetracycline resistance [10] in *E. coli*. We simulated populations of cells over several generations, subjecting each cell to a stochastic environment and providing them the ability to mutate the dynamical properties of this genetic circuit. At the end of each generation, we selected the fittest cells.

We investigated the role and evolvability of both mean expression as well as noise strength of this gene, as key variables in the ability of the cell to cope with changing environmental conditions. We further studied the consequences of subjecting the cells to fluctuating environmental conditions on the genotypic and phenotypic diversity of the cell population over time.

Given an initially homogenous population we found that in stable environments, genotypic diversity is enhanced and then maintained at a given level by neutral mutations that allow the cells to explore various equally beneficial, distinct evolutionary pathways.

Environmental changes were found to be the main enhancer of genotypic diversity, in agreement with observations [7]. When subject to changes in the amounts of toxins its subject to, the cell population not only evolves towards changing appropriately the mean gene expression level, but also towards increasing genotypic diversity in the moments following changes in the external conditions. In some cases, cells evolved both mean expression level, as well as the noise strength, which when increasing causes stronger phenotypic diversity.

We also allowed the evolution of the mutation rates themselves, and found that there are optimal mutation rates whose rate depends on the amount of toxin the cells were subject to. The higher the amount, the higher the mutation rate, since the initial genotype of the cells was proper only for minute amounts of toxin. Higher mutation rates allowed faster reaching of an optimal genotype, at the cost of a higher rate of failure at the individual level, due to harmful mutations.

We conclude that the optimal mutation rates in the model cells depend on both the present level of adaptation of the cells, the necessary degree of change to reach the optimal genotype, as well as the rate of change of environmental conditions when these are unstable. We hypothesize that the ability to generate heritable phenotypic variation [22] as well as the rate of mutation are likely to be evolvable, selectable traits.

Finally, we note that we opted to subject the cells to approximately constant amounts of toxin over long periods of time, and when environmental changes occur, for those changes to be rather "abrupt" in comparison with the total simulation time of the many cell generations. The more abrupt are the changes in expected toxin levels, the more likely it is that a mutation occurring after that change is beneficial in comparison with the previously optimally adapted cells. In the future it will be of interest to test how the rate at which the change occurs affects the results. We expect that the smoother is the environmental change the slower is the selection for mutated cells. However, that will also allow the cell population to maintain at all times, including during the change, a higher mean level of fitness as the cells are more capable of adapting to the changes at the same rhythm as these occur.

To the best of our knowledge, this work is the first view of how a delayed stochastic model of a small genetic circuit may evolve when subject to environmental changes, where the allowed mutations directly affect the kinetics of the genetic circuit, and consequent response to the environment. This was feasible because in the network modeled the protein interacts directly with the toxin. This is not the common scenario, usually there are far more steps between gene expression and interaction with the environment. There are several studies of how perturbations in the environment may affect a population's evolvability, genotypic diversity, etc (see e.g. [23]). However, in general, the processes under evolution are not explicitly modeled. Here we built on these works but, in our model, it is explicitly accounted both the internal stochasticity of the gene's expression dynamics as well as the stochasticity of the environment, and also the stochasticity of the interaction between each cell and its environment. In the future, the use of these models may allow improving our understanding on how the stochasticity in gene expression at the molecular level constrains the evolvability of gene networks.

Acknowledgement. ASR, O-PS, AH, OY-H thank the Academy of Finland (projects 129657 and 126803) and the Finnish Funding Agency for Technology and Innovation (project 40284/08). FGB thanks the Alberta Children's Hospital Research Foundation.

References

1. Kirschner, M., Gerhart, J.: Evolvability. Proc. Natl. Acad. Sci. U.S.A. 95, 8420–8427 (1998)
2. Arkin, A.P., Ross, J., McAdams, H.H.: Stochastic kinetic analysis of developmental pathway bifurcation in phage λ-infected Escherichia coli cells. Genetics 149, 1633–1648 (1998)
3. Samoilov, M., Price, G., Arkin, A.: From fluctuations to phenotypes: The physiology of noise. Science STKE 366, re17 (2006)
4. Kellermayer, R.: Physiologic noise obscures genotype-phenotype correlations. Am. J. Med. Genet. 143A, 1306–1307 (2007)
5. Pal, C., Macia, M., Oliver, A., Schachar, I., Buckling, A.: Coevolution with viruses drives the evolution of bacterial mutation rates. Nature 450, 1079–1081 (2007)
6. Mayr, E.: Variation. In: Hallgrimsson, B., Hall, B.K. (eds.) A Central Concept in Biology. Elsevier Academic Press, Amsterdam (2005)
7. Giraud, A., Matic, I., Tenaillon, O., Clara, A., Radman, M., Fons, M., Taddei, F.: Costs and benefits of high mutation rates: Adaptive evolution of bacteria in the mouse gut. Science 291, 2606–2608 (2001)
8. Suel, G.M., Garcia-Ojalvo, J., Liberman, L.M., Elowitz, M.B.: An excitable gene regulatory circuit induces transient cellular differentiation. Nature 440, 545–550 (2006)
9. Maamar, D., Raj, A., Dubnau, D.: Noise in gene expression determines cell fate in Bacillus subtilis. Science 317, 526–529 (2007)
10. Korpela, M., Kurittu, J., Karvinen, J., Karp, M.: A recombinant Escherichia coli sensor strain for the detection of tetracyclines. Anal. Chem. 70, 4457–4462 (1998)
11. Roussel, M., Zhu, R.: Validation of an algorithm for delay stochastic simulation of transcription and translation in prokaryotic gene expression. Phys. Biol. 3, 274–284 (2006)
12. Gillespie, D.T.: Exact stochastic simulation of coupled chemical reactions. J. Phys. Chem. 81(25), 2340–2361 (1977)
13. Ribeiro, A.S., Lloyd-Price, J.: SGNSim, Stochastic gene networks simulator. Bioinformatics 23(6), 777–779 (2007)
14. Ribeiro, A.S., Zhu, R., Kauffman, S.A.: A General modeling strategy for gene regulatory networks with stochastic dynamics. J. Computational Biology 13(9), 1630–1639 (2006)
15. Yu, J., Xiao, J., Ren, X., Lao, K., Xie, S.: Probing gene expression in live cells, one protein molecule at a time. Science 311, 1600–1603 (2006)
16. Zhu, R., Ribeiro, A.S., Salahub, D., Kauffman, S.A.: Studying genetic regulatory networks at the molecular level: Delayed reaction stochastic models. J. Theoretical Biology 246(4), 725–745 (2007)

17. Eckert, B., Beck, C.F.: Overproduction of transposon Tn10-encoded tetracycline resistance protein results in cell death and loss of membrane potential. J. Bacteriol. 171(6), 3557–3559 (1989)
18. Vogel, U., Jensen, K.: Effects of the antiterminator BoxA on transcription elongation kinetics and ppGpp inhibition of transcription elongation in Escherichia coli. J. Biol. Chem. 270(31), 18335–18340 (1995)
19. Ribeiro, A.S., Hakkinen, A., Mannerstrom, H., Lloyd-Price, J., Yli-Harja, O.: Effects of the promoter open complex formation on gene expression dynamics. Phys. Rev. E 81(1) (2010)
20. Hillen, W., Berens, C.: Mechanisms underlying expression of Tn10 encoded tetracycline resistance. Annu. Rev. Microbiol. 48, 345–369 (1994)
21. Foster, P.L.: Sorting out mutation rates. Proc. Natl. Acad. Sci. U.S.A. 96, 7617–7618 (1999)
22. Gerhart, J., Kirschner, M.: Cells, Embryos, and Evolution. Blackwell Science, Inc., Malden (1997)
23. Maynard Smith, J.: The Evolution of Sex. Cambridge University Press, Cambridge (1978)

Translation from the Quantified Implicit Process Flow Abstraction in SBGN-PD Diagrams to Bio-PEPA Illustrated on the Cholesterol Pathway

Laurence Loewe[1], Maria Luisa Guerriero[1], Steven Watterson[1,2],
Stuart Moodie[3], Peter Ghazal[1,2], and Jane Hillston[1,3]

[1] Centre for System Biology at Edinburgh, King's Buildings,
The University of Edinburgh, Edinburgh EH9 3JD, Scotland
Laurence.Loewe@ed.ac.uk, mguerrie@inf.ed.ac.uk,
[2] Division of Pathway Medicine, The University of Edinburgh
S.Watterson@ed.ac.uk, P.Ghazal@ed.ac.uk
[3] School of Informatics, The University of Edinburgh
Stuart.Moodie@ed.ac.uk, Jane.Hillston@ed.ac.uk

Abstract. For a long time biologists have used visual representations of bio-chemical networks to gain a quick overview of important structural properties. Recently SBGN, the Systems Biology Graphical Notation, has been developed to standardise the way in which such graphical maps are drawn in order to facilitate the exchange of information. Its qualitative Process Description (SBGN-PD) diagrams are based on an implicit Process Flow Abstraction (PFA) that can also be used to construct quantitative representations, which facilitate automated analyses of the system. Here we explicitly describe the PFA that underpins SBGN-PD and define attributes for SBGN-PD glyphs that make it possible to capture the quantitative details of a biochemical reaction network. Such quantitative details can be used to automatically generate an executable model. To facilitate this, we developed a textual representation for SBGN-PD called "SBGNtext" and implemented SBGNtext2BioPEPA, a tool that demonstrates how Bio-PEPA models can be generated automatically from SBGNtext. Bio-PEPA is a process algebra that was designed for implementing quantitative models of concurrent biochemical reaction systems. The scheme developed here is general and can be easily adapted to other output formalisms. To illustrate the intended workflow, we model the metabolic pathway of the cholesterol synthesis. We use this to compute the statin dosage response of the flux through the cholesterol pathway for different concentrations of the enzyme HMGCR that is inhibited by statin.

1 Introduction

Biologists are constantly searching for strategies that help them to understand the complexity of life. Navigating the functional molecular interactions within cells has proven to be an increasing challenge since molecular biological research is filling databases with detailed knowledge about the molecular mechanics of life. A wide variety of schemes has been developed to represent such knowledge, ranging from textual representations that resemble chemical reactions (e.g. Dizzy [42]) or reaction rules (e.g.

C. Priami et al. (Eds.): Trans. on Comput. Syst. Biol. XIII, LNBI 6575, pp. 13–38, 2011.
© Springer-Verlag Berlin Heidelberg 2011

BioNetGen, Kappa [13,12,24]) through XML-based standards like SBML [25] to graphical notations (e.g. [29,43,28,38,15,33]). Graphical maps of biochemical reaction networks are proving to be powerful tools for facilitating an overview of the interactions of particular molecules. Recently the Systems Biology Graphical Notation (SBGN) has emerged as a standard for drawing such reaction diagrams [33,32]. The objective is to provide molecular systems biologists with an easily understandable description of the system by generating consistent maps across different editing tools (e.g. CellDesigner [18], Cytoscape [11], Edinburgh Pathway Editor [46], JDesigner [44]). Like electronic circuit diagrams, they aim to unambiguously describe the structure of a complex network of interactions using graphical symbols.

To achieve this requires a collection of symbols and rules for their valid combination. The SBGN Process Description, SBGN-PD, is a visual language with a precise grammar that builds on an underlying abstraction as the basis of its semantics (see p.40 [33]). We call this underlying abstraction for SBGN-PD the "Process Flow Abstraction" (PFA). It describes biological pathways in terms of processes that transform elements of the pathway from one form into another. The usefulness of an SBGN-PD description critically depends on the faithfulness of the underlying PFA and a tight link between the PFA and the glyphs used in diagrams. The graphical nature of SBGN-PD allows only for qualitative descriptions of biological pathways. However, the underlying PFA is more powerful and also forms the basis for quantitative descriptions that could be used for analysis. Such descriptions, however, need to allow the inclusion of the corresponding mathematical details like parameters and equations for computing the rate at which reactions occur.

Here we aim to make explicit the PFA that already underlies SBGN-PD implicitly. This serves a twofold purpose. First, a better and clearer understanding of the underlying abstraction will make it easier for biologists to construct SBGN-PD diagrams. Second, the PFA is easily quantified and making this explicit can facilitate the quantitative description of SBGN-PD diagrams. Such descriptions can then be used directly for predicting quantitative properties of the system in simulations. Here we demonstrate how this could work by mapping SBGN-PD to a quantitative analysis system. We use the process algebra Bio-PEPA [10,3] as an example, but our mapping can be easily applied to other formalisms as well.

This paper is an extension of previous work presented at the CompMod09 Workshop [36]. Besides small improvements throughout the paper we provide more details on the overall workflow that now includes a working prototype of the Edinburgh Pathway Editor [46] and a fuller introduction to the Bio-PEPA background. Most importantly we apply our toolchain to a completely new example with more entities than the MAPK signalling pathway we used before. As example we now use the metabolic pathway that produces cholesterol, which is modelled in collaboration with colleagues at the Division of Pathway Medicine at the University of Edinburgh. We use our model to investigate how statin inhibits cholesterol production under various circumstances – a question of considerable medical interest [2,8,31].

The rest of the paper is structured as follows. First we provide an overview of the implicit PFA with the help of an analogy to a system of water tanks, pipes and pumps (Section 2). In Section 3 we explain how this system can be extended in order to capture

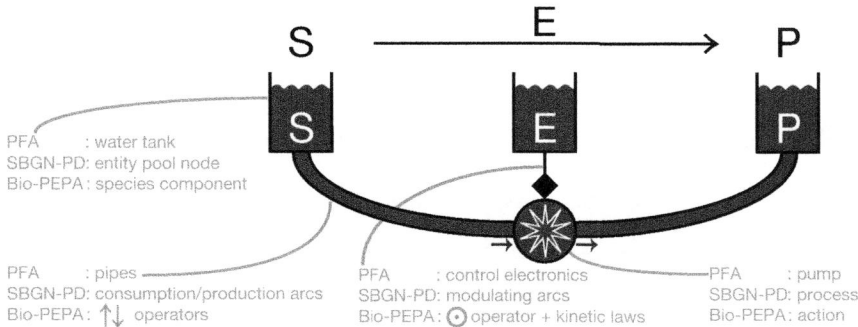

Fig. 1. An overview of the process flow abstraction. The chemical reaction at the top is translated into an analogy of water tanks, pipes and pumps that can be used to visualise the process flow abstraction. The various elements are also mapped into SBGN-PD and Bio-PEPA terminology.

quantitative details of the PFA. We then show how SBGN-PD glyphs can be mapped to a quantitative analysis framework, using the Bio-PEPA modelling environment [3] as an example (Section 4). In Section 5 we discuss various internal mechanisms and data structures needed for translation into any quantitative analysis framework. Section 6 demonstrates the intended workflow by using a model of the cholesterol pathway as an example. We draw a SBGN-PD map of the cholesterol pathway in the Edinburgh Pathway Editor [46] to visualise it and to add quantitative details. The Edinburgh Pathway Editor model can be exported as SBGNtext, which is automatically translated into a Bio-PEPA model by our new translation tool "SBGNtext2BioPEPA" [34,35]. This model is then investigated in the Bio-PEPA Eclipse Plugin. We end by reviewing related work and providing some perspectives for further developments.

2   The Implicit Process Flow Abstraction of SBGN-PD

The PFA behind SBGN-PD is best introduced in terms of an analogy to a system of many water tanks that are connected by pipes. Each pipe either leads to or comes from a pump whose activity is regulated by dedicated electronics. In the analogy, the water is moved between the various tanks by the pumps. In a biochemical reaction system, this corresponds to the biomass that is transformed from one chemical species into another by chemical reactions. SBGN-PD aims to also allow for descriptions at levels above individual chemical reactions. Therefore the water tanks or chemical species are termed "entities" and the pumps or chemical reactions are termed "processes". For an overview, see Figure 1. We now discuss the correlations between the various elements in the analogy and in SBGN-PD in more detail. In this discussion we occasionally allude to SBGNtext, which is a full textual representation of the semantics of SBGN-PD (developed to facilitate automated translation of SBGN-PD into other formalisms; see [34,35]). Here are the key elements of the PFA:

Water tanks = entity pool nodes (EPNs). Each water tank stands for a different pool of entities, where the amount of water in a tank represents the biomass that

Table 1. Categories of "water tanks" in the PFA correspond to types of entity pool nodes in SBGN-PD. The complex and the multimers are shown with exemplary auxiliary units that specify cardinality, potential chemical modifications and other information.

SBGN-PD glyph	EPNType	class type	comment
LABEL	Unspecified	material	EPN (unknown specifics)
LABEL	SimpleChemical	material	EPN
LABEL	Macromolecule	material	EPN
LABEL	NucleicAcidFeature	material	EPN
LABEL LABEL LABEL	-	material	EPN multimer specified by cardinality
LABEL LABEL LABEL	Complex	container	EPN, arbitrary nesting
∅	Source	conceptual	external source of molecules
∅	Sink	conceptual	removal from the system
⟩LABEL⟨	PerturbingAgent	conceptual	external influence on a reaction

is bound in all entities of that particular type that exist in the system. Typical examples for such pools of identical entities are chemical species like metabolites or proteins. SBGN-PD does not distinguish individual molecules within pools of entities, as long as they are within the same compartment and identical in all other important properties. An overview of all types of EPNs (i.e. categories of water tanks) in SBGN-PD is given in Table 1. To unambiguously identify an entity pool in SBGNtext and in the code produced for quantitative analysis, each entity pool is given a unique `EntityPoolNodeID`. The PFA does not conceptually distinguish between non-composed entities and entities that are complexes of other entities. Despite potentially huge differences in complexity they are all "water tanks" and further quantitative treatment does not treat them differently.

Pipes = consumption and production arcs. Pipes allow the transfer of water from one tank to another. Similarly, to move biomass from one entity pool to another requires the consumption and production of entities as symbolised by the corresponding arcs in SBGN-PD (see Table 3). These arcs connect exactly one process and one EPN. The thickness of the pipes could be taken to reflect stoichiometry, which is the only explicit quantitative property that is an integral part of SBGN-PD. Production arcs take on a special role in reversible processes by allowing for bidirectional flow.

Pumps = processes. Pumps move water through the pipes from one tank to another. Similarly, processes transform biomass bound in one entity to biomass bound in another entity, i.e. processes transform one entity into another. The speed of the pump in the analogy corresponds to the frequency with which the reaction occurs and determines the amount of water (or biomass) that is transported between tanks (or that is converted from one entity to another, respectively). Processes can belong to different types in SBGN-PD (Table 2) and are unambiguously identified by a unique ProcessNodeID in SBGNtext. This allows arcs to clearly define which process they belong to and, by finding all its arcs, each process can also identify all EPNs it is connected to.

Reversible processes. SBGN-PD allows for processes to be reversible if they are symmetrically modulated (p.28 [33]). Thus, there may be flows in two directions. However the net flow at any given time will be unidirectional. The PFA does not prescribe how to implement this. For simplicity, our analogy assumes pumps to be unidirectional, like many real-world pumps. Thus bidirectional processes in our analogy are represented as two pumps with corresponding sets of pipes and opposite directions of flow. In a reversible process the products of the forward process are consumed in the backward process, thus Consumption and Production arcs can no longer be as clearly separated as in unidirectional processes. To resolve this, SBGN-PD distinguishes the left-hand side from the right-hand side of a process and uses only arcs that look like Production arcs to indicate the double role (p.32 [33]). In SBGN-PD reversible process nodes are easy to recognise visually by the absence of Consumption arcs on both sides. To represent all such arcs either as Consumption arcs or as Production arcs in SBGNtext would lose the information of which arc is on which side of the process node. Thus we define two new arc types that are only used for products and reactants in the context of reversible processes: LeftHandSide and RightHandSide. LeftHandSide arcs indicate that they are consumption arcs in the forward process (and production arcs in the backward process), where as RightHandSide arcs are the corresponding opposite. To support reversible processes the visual editor needs to identify reversible processes and assign the corresponding arc types LeftHandSide and RightHandSide to the arcs. In addition a forward and a backward kinetic law need to be stored to facilitate breaking up a bidirectional process into two unidirectional processes.

Control electronics for pumps = modulating arcs and logic gates. In the analogy, pumps need to be regulated, especially in complex settings. This is achieved by control electronics. In SBGN-PD, the same is done by various types of modulation arcs, logic arcs and logic gates [33]. They all contribute to determining the frequency of the reaction. Since SBGN-PD does not quantify these interactions, most of our extensions for quantifying SBGN-PD address this aspect. Each arc connects a "water tank" with a given EntityPoolNodeID and a "pump" with a given ProcessNodeID. Ordinary modulating arcs can be of type Modulation (most generic influence on reaction), Stimulation (catalysis or positive allosteric regulation), Catalysis (special case of stimulation, where activation energy is lowered), Inhibition (competitive or allosteric) or NecessaryStimulation (process is only possible if the stimulation is "active", i.e. has surpassed some threshold). The glyphs are shown in Table 3, where their mapping to Bio-PEPA is discussed. One might misread SBGN-PD to suggest that Consumption / Production arcs cannot modulate the frequency of a process. However, kinetic

Table 2. Categories of "pumps" in the process flow abstraction correspond to types of processes in SBGN-PD. The grey lines indicate that more than one EPN can participate in this process.

SBGN-PD glyph	ProcessType	meaning
>—□—<	Process	normal known processes
>—●—	Association	special process that builds complexes
—◎—<	Dissociation	special process that dissolves complexes
>—◊—<	Omitted	several known processes are abstracted
>—?—<	Uncertain	existence of this process is not clear
⟨LABEL⟩	Observable	this process is easily observable

laws frequently depend on the concentration of reactants, implying that these arcs can also contribute to the "control electronics" (e.g. report "level of water in tank"). Another part of the "control electronics" are *logical operators*. These simplify modelling, when a biological function can be approximated by a simple on/off logic that can be represented by boolean operators. SBGN-PD supports this simplification by providing the logical operators "AND", "OR" and "NOT". These take "logic arcs" as input and output, which convert a molecule count into a digital signal and back.

Groups of water tanks = compartments, submaps and more. The PFA is complete with all the elements presented above. However, to make SBGN-PD more useful for modelling in a biological context, SBGN-PD has several features that make it easier for biologists to recognise various subsets of entities that are related to each other. For example, entities that belong to the same compartment can be grouped together in the compartment glyph and functionally related entities can be placed on the same submap. In the analogy, this corresponds to grouping related water tanks together. SBGN-PD also supports sophisticated ways for highlighting the inner similarities between entities based on a knowledge of their chemical structure (e.g. modification of a residue, formation of a complex). Stretching the analogy, this corresponds to a way of highlighting some similarities between different water tanks. These groupings are only conceptual and have no effect on quantitative analysis, as long as different "water tanks" remain separate.

3   Extensions for Quantitative Analysis

The process flow abstraction that is implicit in all SBGN process diagrams can be used as a basis to quantify the systems they describe. Following the introduction to the PFA above, we now discuss the attributes that need to be added to the various SBGN-PD glyphs in order to allow for automatic translation of SBGN-PD diagrams into quantitative models. These attributes are stored as strings in SBGNtext (our textual representation of SBGN-PD, see [35]) and are attached to the corresponding glyphs by a graphical

SBGN-PD editor. They do not require a visual representation that compromises the visual ease-of-use that SBGN-PD aims for. A prototypic example of how the quantitative information could be added in a visual editor is provided by the Edinburgh Pathway Editor [46] and shown in Figure 2. Next we discuss the various attributes that are necessary for the glyphs of SBGN-PD to support quantitative analysis. We do not discuss SBGN-PD glyphs for auxiliary units, submaps, tags and equivalence arcs here, as they do not require extensions for supporting quantitative analysis.

3.1   Quantitative Extensions of EntityPoolNodes

For quantitative analysis, each unique EPN requires an `InitialMoleculeCount` to unambiguously define how many entities exist in this pool in the initial state. We followed developments in the SBML standard in using counts of molecules instead of concentrations, since SBGN-PD also allows for multiple compartments, making the use of concentrations very cumbersome (see section 4.13.6, p.71f. in [25]). For entities of type `Perturbation`, the `InitialMoleculeCount` is interpreted as the numerical value associated with the perturbation, even though its technical meaning is not a count of molecules. Entities of the type `Source` or `Sink` are both assumed to be effectively unlimited, so `InitialMoleculeCount` does not have a meaning for these entities. Beyond a unique `EntityPoolNodeID` and `InitialMoleculeCount`, no other information on entities is required for quantitative analysis.

3.2   Quantitative Extensions of Arcs

Arcs link entities and processes by storing their respective IDs and the `ArcType`. The simplest arcs are of type `Consumption` or `Production` and do not require numerical information beyond the stoichiometry that is already defined in SBGN-PD as a property of arcs that can be displayed visually in standard SBGN-PD editors. Logic arcs will be discussed below. All modulating arcs are part of the "control electronics" and affect the frequency with which a process happens. They link to EPNs to inform the process about the presence of enzymes, for example. Modulation is usually governed by parameters or other important quantities for the given process (e.g. Michaelis-Menten constant).

To make the practical encoding of a model easier, we define process parameters that conceptually belong to a particular modulating entity as a list of `QuantitativeProperties` in the arc pointing to that entity. This is equivalent to seeing the set of parameters of a reaction as something that is specific to the interaction between a particular modulator and the process it modulates. Other approaches are also possible, but lead to less elegant implementations. Storing parameters in equations requires frequent and possibly error-prone changes (e.g. many different Michaelis-Menten equations). One could also argue that the catalytic features are a property of the enzyme and thus make parameters part of EPNs; however this would mean that all the reactions catalysed by the same enzyme would have the same parameters or would require cumbersome naming conventions to manage different affinities for different substrates.

To refer to parameters we specify the `ManualEquationArcID` of an arc and then the name of the parameter that is stored in the list of `QuantitativeProperties` of that arc. This scheme reduces clutter by limiting the scope of the relevant namespace (only few arcs per process exist, so `ManualEquationArcIDs` only need to be unique within

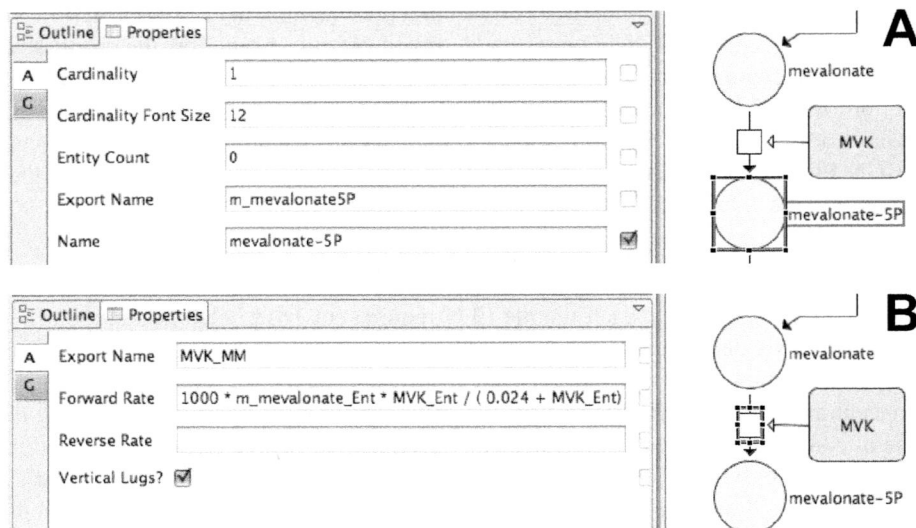

Fig. 2. An example of how attributes attached to SBGN-PD glyphs and stored as strings can be used to add quantitative information to a visual representation of a biochemical reaction network. These screenshots from the Edinburgh Pathway Editor (Version 3.0.0-alpha13) [46] show a selected glyph with its attributes that are automatically displayed in the properties window. (A) EntityPoolNode "Entity Count" is mapped to `InitialMoleculeCount`. (B) ProcessNode with attributes for entering the propensity functions for the forward and backward reactions. "Export Name" facilitates the production of readable Bio-PEPA models.

that immediate neighbourhood). Thus parameter names can be brief, since they only need to be unique within the arc. The `ManualEquationArcID` is specified by the user in the visual SBGN-PD editor and differs from ArcID, a globally unique identifier that is automatically generated by the graphical editor. The `ManualEquationArcID` allows for user-defined generic names that are easy to remember, such as "Km" and "vm" for Michaelis-Menten reactions. It should be easily accessible within the graphical editor, just as the parameters that are stored within an arc.

Logical operators and logic arcs. To facilitate the use of logical operators in quantitative analyses one needs to convert the integer molecule counts of the involved EPNs to binary signals amenable to boolean logic. Thus SBGNtext supports "incoming logic arcs" that connect a "source entity" or "source logical result" with a "destination logic operator" and apply an "input threshold" to decide whether the source is above the threshold ("On") or below the threshold ("Off"). To this end, a graphical editor needs to support the "input threshold" as a numerical attribute that the user can enter; all other information recorded in incoming logic arcs is already part of an SBGN diagram. Once all signals are boolean, they can be processed by one or several logical operators, until the result of this operation is given in the form of either 0 ("Off") or 1 ("On"). This result then needs to be converted back to an integer or float value that can be further processed to compute process frequencies. Thus a graphical editor needs to support corresponding attributes for defining a low and a high output level.

3.3   Quantitative Extensions of ProcessNodes

For quantitative analyses, a ProcessNode must have a unique name and a kinetic law that represents the propensity, which is proportional to the probability that this process occurs next in a stochastic model, based on the current global state of the model. In a deterministic model this equation gives a rate law that is expressed in terms of absolute molecule numbers, not concentrations. Since the `ProcessType` is not required for quantitative analyses, it does not matter whether a process is an ordinary `Process`, an `Uncertain` process or an `Observable` process, for example. For all these ProcessNodes, graphical editors need to support attributes for the manual specification of a `ProcessNodeID`, and a `PropensityFunction`. These attributes are then stored in SBGNtext. If support for bidirectional processes is desired, then graphical editors need to facilitate entering a propensity function for the backward process as well. Propensity functions compute the propensity of a unidirectional process to be the next event in the model and can be used directly by simulation algorithms and ODE solvers [20].

A `PropensityFunction` can be given directly (see current prototype of Edinburgh Pathway Editor [46]; Figure 2), but the full definition of SBGNtext specifies propensities by referring to aliases. This can simplify the specification of models and hence reduce errors. For instantiation, a translator needs to replace all aliases by their true identity. We use the following syntax for a parameter alias that is substituted by the actual numeric value (or a globally defined parameter) from the corresponding arc:

> `<par: ManualEquationArcID.QuantitativePropertyName>`

While translating to Bio-PEPA this would be simply substituted with a corresponding parameter name. The parameter is then defined elsewhere in the Bio-PEPA model to have the numerical value stored in the corresponding property of the arc. To allow the numerical analysis tool to access an EPN count at runtime we replace the following entity alias by the `EntityPoolNodeID` that the corresponding arc links to:

> `<ent: ManualEquationArcID>`

This is shorter than the `EntityPoolNodeID` and allows the reuse of propensity functions if kinetic laws are identical and the manual IDs follow the same pattern. It is desirable that there is no need to specify the `EntityPoolNodeID` since it is fairly long and generated automatically to reflect various properties that make it unique. It would be cumbersome to refer to in the equation and it would require a mechanism to access the automatically generated `EntityPoolNodeID` before a SBGNtext file is generated. Also any changes to an entity that would affect its `EntityPoolNodeID` would then also require a change in all corresponding propensity functions, a potentially error-prone process. The same substitution mechanism can be used to provide access to properties of compartments (see [35]). In addition to these aliases, functions use the typical standard arithmetic rules and operators that are directly passed through to the next level.

4   Mapping SBGN-PD Elements to Bio-PEPA

In this section we explain how to use the semantics of SBGN-PD to map a SBGN-PD model to a formalism for quantitative analysis. We are using Bio-PEPA as an example, but our approach is general and can be applied to many other formalisms that support the modelling of chemical reactions.

4.1 The Bio-PEPA Language

Bio-PEPA is a stochastic process algebra which models biochemical pathways as interactions of distinct entities representing reactions of chemical species [10,3]. A process algebra model captures the behaviour of a system as the actions and interactions between a number of entities, where the latter are often termed "processes", "agents" or "components". In PEPA [23] and Bio-PEPA [10] these are built up from simple *sequential components*. Different process algebras support different modelling styles for biochemical systems [5]. Stochastic process algebras, such as PEPA [23] or the stochastic π-calculus [41], associate a random variable with each action to represent the mean of its exponentially distributed waiting time. In the stochastic π-calculus, interactions are strictly binary whereas in Bio-PEPA the more general multiway synchronisation is supported. Bio-PEPA is based on the following underlying principles (see [10] for more details):

- modelling follows the "reagent-centric" style, which means that different species components denote different types of reagents;
- only irreversible reactions are considered: reversible reactions can be seen as the union of a pair of forward and backward reactions;
- the reactants of the reaction can only decrease their concentration, the products can only increase it, whereas enzymes and inhibitors do not change;
- a single species in different states (e.g. phosphorylated, free, bound ligand, in different compartments, ...) is regarded as different species and represented by distinct sequential components;
- compartments are static and do not play an active role in reactions, but they can be used to constrain reaction occurrences to a particular location and propensity functions can depend on their size. Here for the sake of simplicity, we assume all species are located in the same compartment.

The syntax of Bio-PEPA is defined as [10] :

$$S ::= (\alpha, \kappa) \text{ op } S \mid S + S \mid C \qquad P ::= P \bowtie_{\mathcal{L}} P \mid S(x)$$

where S is a sequential *species component* that represents a chemical species (termed "process" in some other process algebras and "EntityPoolNode" in SBGN-PD), C is a name referring to a species component defined as $C \equiv S$, P is a *model component* that describes the set $\mathcal{L}$ of possible interactions between species components (these "interactions" or "actions" correspond to "processes" in SBGN-PD and can represent chemical reactions). An initial count of molecules or a concentration of S is given by $x \in \mathbb{R}_0^+$. In the prefix term "$(\alpha, \kappa) \text{ op } S$", κ is the *stoichiometry coefficient* and the operator op indicates the role of the species in the reaction α. Specifically, op $= \downarrow$ denotes a *reactant*, $\uparrow$ a *product*, $\oplus$ an *activator*, $\ominus$ an *inhibitor* and $\odot$ a generic *modifier*, which indicates more generic roles than $\oplus$ or $\ominus$. The operator "+" expresses a choice between possible actions. Finally, the process $P \bowtie_{\mathcal{L}} Q$ denotes the synchronisation between components:

the set $\mathcal{L}$ determines those activities on which the operands are forced to synchronise. When $\mathcal{L}$ is the set of common actions, we use the shorthand notation $P \bowtie Q$. A Bio-PEPA model $\mathcal{P}$ is defined as a 6-tuple $\langle \mathcal{V}, \mathcal{N}, \mathcal{K}, \mathcal{F}_R, Comp, P \rangle$, where: $\mathcal{V}$ is the set of compartments, $\mathcal{N}$ is the set of quantities describing each species, $\mathcal{K}$ is the set of all parameters referenced elsewhere, $\mathcal{F}_R$ is the set of functional rates that define all required kinetic laws, $Comp$ is the set of definitions of species components S that highlight the reactions a species can take part in and P is the system model component.

A variety of analysis techniques can be applied to a single Bio-PEPA model, facilitating the easy validation of analysis results when the analyses address the same issues [4] and enhancing insight when the analyses are complementary [9,1]. Currently supported analysis techniques include stochastic simulation at the molecular level, ordinary differential equations, probabilistic and statistical model-checking and numerical analysis of continuous time Markov chains [10,3,17]. Additional analysis techniques are facilitated by compositional reasoning, which allows the automated extension of elementary proofs of qualitative features to complex models. Examples for such qualitative analyses include deadlock and livelock detection and model-checking of a model against a logical formula.

4.2   SBGN-PD Mapping

Here we map the core elements of SBGN-PD to Bio-PEPA (see [34] for an implementation).

Entity Pool Nodes. Due to the rich encoding of information in the `EntityPoolNode-ID`, Bio-PEPA can treat each distinct `EntityPoolNodeID` as a distinct species component. This removes the need to explicitly consider any other aspects such as entity type, modifications, complex structures and compartments, as all such information is implicitly passed on to Bio-PEPA by using the `EntityPoolNodeID` as the name for the corresponding species component. The definition of the set $\mathcal{N}$ of a Bio-PEPA system requires the attribute `InitialMoleculeCount` for each EPN (see Section 3).

Processes. All SBGN-PD `ProcessTypes` are represented as reactions in Bio-PEPA. Compiling the corresponding set $\mathcal{F}_R$ relies on the attribute `PropensityFunction` and a substitution mechanism that makes it easy to define these functions manually. To help humans understand references to processes in the sets $\mathcal{F}_R$ and $Comp$ requires recognisable names for SBGN-PD `ProcessNodeIDs` that map directly to their identifiers in Bio-PEPA. Thus graphical editors need to support manual `ProcessNodeIDs`.

Reversible processes. The translator supports reversible SBGN-PD processes by dividing them into two unidirectional processes for Bio-PEPA. The translator reuses the manually assigned `ProcessNodeID` and augments it by "_F" for forward reactions and "_B" for backward reactions. These two unidirectional processes are then treated independently. When compiling the species components in Bio-PEPA, every time a `LeftHandSide` arc is found, the translator assumes that the corresponding forward and backward processes have been defined and will augment the process name appropriately, while adding the corresponding Bio-PEPA operator for reactant and product.

Table 3. "Water pipes and control electronics": Mapping arcs between entities and processes in SBGN-PD to operators in Bio-PEPA species components. "Symbols" are the formal syntax of Bio-PEPA, while "code" gives the concrete syntax used in the Bio-PEPA Eclipse Plug-in [3].

SBGN-PD glyph	ArcType	Bio-PEPA symbol	Bio-PEPA code
Origin EPN — N	Consumption	$\downarrow$	<<
N ▶ Target EPN	Production	$\uparrow$	>>
N ▶ Target EPN	LeftHandSide	$\downarrow$ and $\uparrow$	<< and >>
N ▶ Target EPN	RightHandSide	$\uparrow$ and $\downarrow$	>> and <<
—◇ Target TRN	Modulation	$\odot$	(.)
—▷ Target TRN	Stimulation	$\oplus$	(+)
—○ Target TRN	Catalysis	$\oplus$	(+)
—⊣ Target TRN	Inhibition	$\ominus$	(-)
—▮▷ Target TRN	NecessaryStimulation	$\odot$	(.)

RightHandSide arcs are handled in the same way. Thus the production arc glyph in SBGN-PD has three distinct meanings as shown in Table 3.

Arcs. The arcs in SBGN-PD define which entities participate in which processes. Thus arcs play a pivotal role in defining the species components in Bio-PEPA. Since arcs can store kinetic parameters, they are also important for defining parameters in Bio-PEPA. As kinetic law definitions in Bio-PEPA frequently refer to such parameters, we use the ArcID that is automatically generated by the graphical editor to substitute the local manual arc references in propensity functions by globally unique parameters names (see Section 3). The type of an arc indicates both the role of the connected entity in the process (consumed reactant, product or rate modifier) and the chemical nature of the reaction (catalysis, stimulation, inhibition, necessary stimulation or the most generic modification). Thus the type of an arc can be mapped directly to the operator "op" described in the Bio-PEPA syntax shown in Table 3. All mappings are straightforward except NecessaryStimulation (previously called Trigger), which we mapped to the generic modifier $\odot$ to indicate that this interaction inhibits below and stimulates above a given threshold.

Logical operators. Logical operators require the conversion of integer molecule counts of the relevant EPNs to binary signals and after some boolean logic processing back to low and high integer values. As evident from the implementation scheme above, the use of all quantitative properties culminates in the correct formulation of the corresponding propensity functions that determine the probability that the corresponding process will be the next to occur. Thus an implementation of logical operators requires that their results be included in the corresponding propensity functions. The current scheme of implementing propensity functions relies heavily on substituting the various components into the final equation, so that Bio-PEPA will ultimately only see one formula per propensity function. In this context the implementation of logical operators requires the insertion of a formula in the propensity function that computes the result of the boolean operations from their integer input. An arbitrarily complex logic operator network can be constructed from the following basic building blocks:

- *Convert from integers or double floats to boolean values.* This is best done by a specially defined mathematical function that takes an integer or float signal and compares it to a specified threshold, returning either 0 (signal $\leq$ threshold) or 1 (signal > threshold). The definition of such a function is not complicated and can be implemented with the help of the Heaviside step function that is available in the Bio-PEPA Eclipse Plug-in.
- *Map boolean operators:* AND $\rightarrow$ multiply all boolean inputs to get output. NOT $\rightarrow$ use the arithmetic expression *output* = (1 − *input*). OR $\rightarrow$ sum all inputs (0/1) and test if it is greater than 1 using the threshold function.

5   Converter Implementation and Internal Representation

We chose Java as implementation language for the converter described above, due to the good portability of the resulting binaries and for interoperability with the Bio-PEPA Eclipse Plug-in [3]. We defined a grammar for SBGNtext in the Extended Backus-Naur-Form (EBNF) as supported by ANTLR [39], which automatically compiles the Java sources for the parser that stores all important parsing results in a number of co-herently organised internal TreeMaps. To generate a Bio-PEPA model three main loops over these TreeMaps are necessary: over all entities, over all processes and over all parameters. To illustrate the translation we refer to "code" examples from Figure 3.

The **loop over all entities** (e.g. "m_LSS_Ent") compiles the species components as well as the model component required by Bio-PEPA. The latter is a list of all partici-pating `EntityPoolNodeIDs` combined by the cooperation operator "$< * >$" or $\bowtie_*$ that automatically synchronises on all common actions ("*"). This simplification depends on all processes in SBGN-PD having unique names and fixed lists of reactants with no mutually exclusive alternatives in them. The first condition can be enforced by the tools that produce the code, the second is ensured by the reaction-style of describing processes in SBGN-PD. For example, SBGN-PD does not allow for a *single* reaction called "bind", which states that A binds with *either* B *or* C to produce D. In Bio-PEPA these alternative reactions could be given the same name and careful construction of the model equation could then ensure that only one of B or C participates in any one occurrence of the reaction. To describe the same model in SBGN-PD requires two reac-tions with different names (A+B→D; A+C→D;). This is then translated into the correct Bio-PEPA model using only "$< * >$". Hence individual actions synchronised by the cooperation operator do not need to be tracked in this system.

For each species component a loop over all arcs finds the arcs that are connected to it (e.g. "st33") and that store all relevant `ProcessNodeIDs` (e.g. "LSS_Proc"). The same loop determines the respective role of the component (as reflected by the choice of the Bio-PEPA operator in Table 3; e.g. "(+)"). To compile this we loop over all arcs to find the arcs that connect to a particular entity. Since the arc also contains the

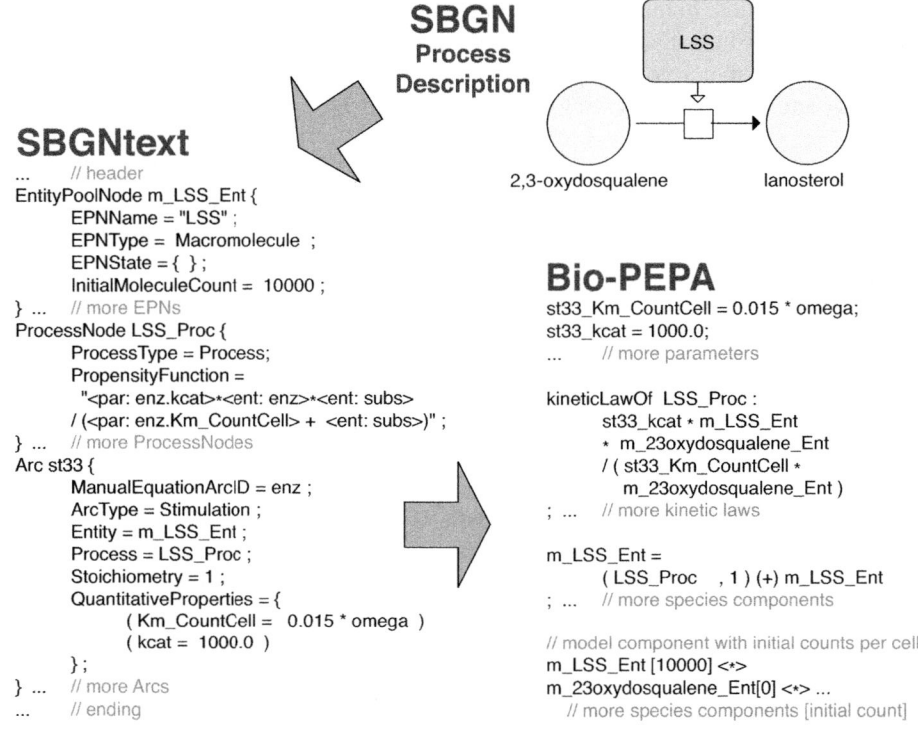

Fig. 3. Transformation of biochemical reaction systems in the automated workflow from SBGN-PD to Bio-PEPA. Using an extract of the cholesterol pathway model discussed below, this example shows how parts from one representation are transformed into parts of another. The code excerpts focus on the enzyme ("LSS_Proc") and the reaction it catalyses ("LSS_Proc"). Parameters such as kinetic constants and initial amounts of species must be scaled from the start in the Edinburgh Pathway Editor to represent discrete molecule counts.

process ID and no further information about a process is required here, there is no need to loop over all processes as well. By combining the information stored in the TreeMaps of entities and arcs, it is possible to compile the relevant information for Bio-PEPA species components. The fact that multiple arcs can connect the same entity to multiple reactions and the same reaction to multiple entities facilitates the preservation of SBGNs capacity to model reactions with many entities during the translation process.

The **loop over all processes** (e.g. "LSS_Proc") compiles the kinetic laws by substituting aliases (e.g. "<ent: enz>" or "<par: enz.kcat>") for EPNs ("m_LSS_Ent") and parameters ("st33_kcat") in the propensity functions specified in the graphical editor. Each function is handled separately by a dedicated function parser that queries the TreeMaps generated when parsing the SBGNtext file.[1]

---

[1] Our prototypic implementation of the workflow presented here passes propensity functions entered in the Edinburgh Pathway Editor without modification to Bio-PEPA. A version of SBGNtext2BioPEPA that substitutes parameters in functions is available from [34].

The **loop over all quantitative properties** of the model defines the parameters in Bio-PEPA (e.g. "st33_kcat"). It is possible to avoid this step by inserting the direct numerical values into the equations processed in the second loop. However, this substantially reduces the readability of equations in the Bio-PEPA model and makes it difficult for third party tools to assist in the automated generation of parameter combinations. Thus we defined a scheme that automatically generates parameter names to maximise the readability of equations (combine ArcID "st33" and name of the quantitative property "kcat").

To facilitate walking through the various collections specified above, the TreeMaps are organised in four sets, one set for entities, one for processes, one for arcs and one for quantitative properties. Each of these sets is characterised by a common key to all maps within the set. This facilitates the retrieval of related parse products for the same key from a different map. Since all this information is accessible from Java code, it is easily conceivable to use the sources produced in this work for reading SBGNtext files in a wide variety of contexts. The system is easy to deploy, since it involves few files and the highly portable ANTLR runtime library. Our converter is called SBGNtext2BioPEPA and sources are available [34].

6   Example: Competitive Inhibition in the Cholesterol Pathway

6.1   A Model of the Cholesterol Synthesis Pathway

Here we illustrate our translation workflow by using the cholesterol synthesis pathway as an example. We first draw an SBGN-PD map of this pathway (Figure 4) in the Edinburgh Pathway Editor (EPE) [46] based on the biochemical reactions listed in KEGG [27]. Then we add the necessary quantitative extensions as attributes to the corresponding glyphs in EPE (for screenshot see Figure 2), before we export the model from EPE to SBGNtext. SBGNtext is automatically translated to Bio-PEPA with the help of SBGNtext2BioPEPA [34]. Finally the model is simulated in the Bio-PEPA Eclipse Plugin Version 0.1.7 [3] using the Gibson-Bruck stochastic simulation method [19] and the Adaptive Dormant Prince ODE solver. Fluxes are computed from the exported time courses as described below. Since our model is focussed on the flux of *de novo* cholesterol production, we choose to ignore the complex processes that degrade or store cholesterol.

Scaling of the System. Since Bio-PEPA and SBML [25] describe systems in terms of explicit molecule counts and not concentrations, we introduce a scaling factor Ω which is used to represent the size of the system. The factor Ω effectively converts a concentration [mM] into a molecular count by multiplying it with Avogadro's number and a volume. As a volume we cannot use the typical volume of a cell, since cholesterol synthesis is confined to the endoplasmatic reticulum, which comprises only a fraction of the cell. At this stage we do not have information about the volume of a cell dedicated to cholesterol synthesis. However, rough estimates show that many typical enzymes that are not produced in particularly high copy numbers exist in about 10^4 copies / cell. Thus we

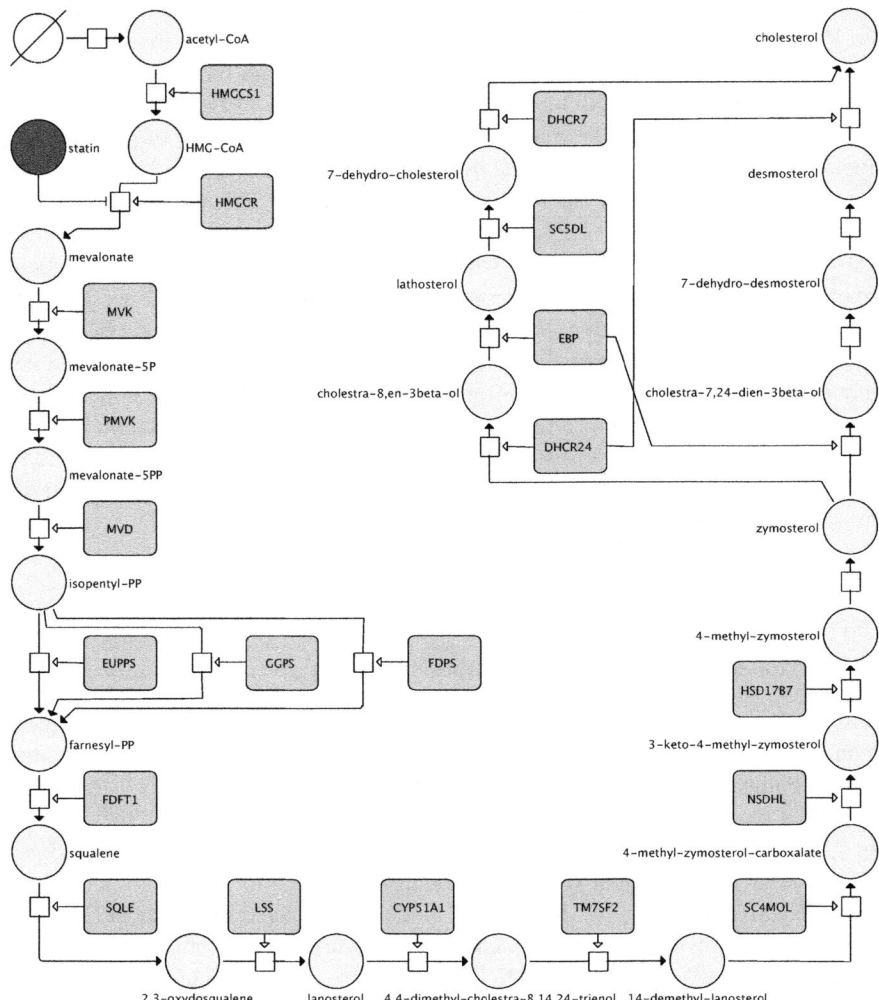

Fig. 4. A SBGN-PD representation of the cholesterol pathway as taken from KEGG [27], drawn in the Edinburgh Pathway Editor [46]. Metabolites are marked as yellow to highlight the mostly linear structure of the pathway. Enzymes are coloured in cyan and the inhibitory statin in red. A constant flow of acetyl-CoA is assumed to enter the system.

choose Ω such that a 10 mM enzyme concentration translates into 10000 copies / cell ($\Omega = 1000$), giving our model a realistic scale. Explicitly representing Ω increases the flexibility of the model by allowing quick changes to the size of the system[2].

---

[2] Since parameters cannot be defined explicitly in the current EPE prototype, we model Ω as a "dummy-species", that does not take part in reactions, but has a constant value, which can be referred to in propensity functions for the purpose of scaling K_M. Since the initial "Entity Count" needs to be an integer of the right size we enter 10000 directly ($= 10 \cdot \Omega$).

Table 4. Kinetic parameters for Michaelis-Menten reactions as used in our model. Values for the reaction rate parameters were taken from BRENDA [7] where possible. In order to have units uniformly in terms of molecule counts, K_M needs to be multiplied with the system size Ω that is also used to specify the enzyme count (see as in Eq.1). Values marked with "*" are hypothetical. Reactions not listed here are assumed to follow mass action kinetics with rate constants of 1000 (reactants: 4-methyl-zymosterol, cholestra-7,24-dien-3β-ol, 7-dehydro-desmosterol) or {0.05, 50} × 10 Ω (product: acetyl-CoA).

Enzyme	Enzyme count	Turnover k_{cat} [1/h]	K_M [mM]
HMGCS1	10 Ω *	1000 *	0.01
HMGCR	{3, 10, 30} Ω *	500 *	0.07
MVK	10 Ω *	1000 *	0.024
PMVK	10 Ω *	36720	0.025
MVD	10 Ω *	17640	0.0074
EUPPS	10 Ω *	1000 *	0.01 *
GGPS	10 Ω *	1000 *	0.01 *
FDPS	10 Ω *	1000 *	0.01 *
FDFT1	10 Ω *	1908	0.0023
SQLE	1000 Ω *	65.88	0.0077
LSS	10 Ω *	1000 *	0.015
CYPY1A1	10 Ω *	1000 *	0.005
TM7SF2	10 Ω *	1000 *	0.0333
SC4MOL	10 Ω *	1000 *	0.01 *
NSDHL	10 Ω *	1000 *	0.007
HSD17B7	1000 Ω *	177.48	0.236
DHCR24 - zymosterol	10 Ω *	1000 *	0.037
DHCR24 - desmosterol	10 Ω *	1000 *	0.01 *
EBP - cholestra-8,en3β-ol	10 Ω *	5122.8	0.01 *
EBP - zymosterol	10 Ω *	1522.8	0.05
SC5DL	10 Ω *	1000 *	0.032
DHCR7	10 Ω *	1000 *	0.277

Reaction Kinetics. Our typical kinetic law (propensity function for stochastic simulations; rate law for ODE) for standard Michaelis Menten kinetics as entered in EPE is

$$\frac{k_{cat} \cdot EnzymeName\_Ent \cdot m\_SubstrateName\_Ent}{(K_M \cdot \Omega) + m\_SubstrateName\_Ent} \quad (1)$$

This function is used to model all reactions that involve one enzyme and one metabolite. To do this we retrieved kinetic parameter values from BRENDA [7]. If no kinetic parameter was found for an enzyme, we assumed a turnover of $k_{cat}=1000$ [1 / h] and a Michaelis-Menten constant of $K_M=0.01$ [mM], which is of the same order of magnitude as the mean of corresponding values of other enzymes that have been observed in experiments. Table 4 reports all relevant parameters used in our model. The reaction catalysed by HMGCR has been described as a rate limiting step of the cholesterol pathway [30,37]. To capture this notion in our model we increase the enzyme copy numbers for SQLE and HSD17B7, two slow reactions for which both k_{cat} and K_M are known.

For reactions without an enzyme in Figure 4 we assumed mass action kinetics. We chose rates so that they would not be limiting in this system and would not force the accumulation of large amounts of reactants (see Table 4).

The first mass action reaction is special since it determines the flux of acetyl-CoA into the system. In the absence of degradation reactions for intermediate metabolites and the endproduct cholesterol, the influx determines the rate of cholesterol production – unless intermediate metabolites accumulate (see below). We chose to model two scenarios: a low-flux scenario that represents situations where acetyl-CoA in the cell is directed away from cholesterol synthesis and a high-flux scenario that reflects conditions of more abundant acetyl-CoA. In the low and high flux scenarios $500 = 0.05 \cdot 10 \, \Omega$ and $500000 = 50 \cdot 10 \, \Omega$ molecules of acetyl-CoA are introduced into the system, respectively ($10 \, \Omega$ was chosen to be of the same order of magnitude as the typical number of copies per enzyme).

In order to model competitive inhibition by statin we use the following kinetic law:

$$\frac{k_{cat} \cdot EnzymeName\_Ent \cdot m\_SubstrateName\_Ent}{m\_SubstrateName\_Ent + (K_M \cdot \Omega) \cdot (1 + m\_InhibitorName\_Ent/(\Omega \cdot K_I))} \quad (2)$$

where HMGCR is the enzyme, HMG-CoA the metabolite, statin the inhibitor and $K_I = 0.000044$ [mM] the inhibition constant average of 11 values for human cells found in BRENDA; values for other reactions are given in Table 4 . From the nature of this function it follows that any number of inhibitor molecules can be rendered ineffective, if countered by a sufficiently large number of metabolites (see results below).

Measurement of Cholesterol Flux. Due to the linear structure of the pathway, its unidirectional flow and the lack of degradation of intermediate products, a steady flow of acetyl-CoA leads to a steady production of cholesterol. If the rate of an intermediate reaction is too low, the production of cholesterol is slowed down temporarily until the substrate of the reaction has accumulated enough to compensate for the reduction. A simulation in the high flux environment without statin and for HMGCR = 30000 showed that all intermediates equilibrate after less than 5 minutes and then fluctuate around fairly low molecule counts (most around zero, all below 500). To facilitate measurements of cholesterol production flux F, we omitted cholesterol degrading reactions. Instead we compute F, the flux of newly synthesised molecules of cholesterol / hour as

$$F = \frac{C_{T_2} - C_{T_1}}{T_2 - T_1} \quad (3)$$

where C represents accumulated counts of synthesised cholesterol molecules in the system at the corresponding points T_1 and T_2 and time is measured in hours.

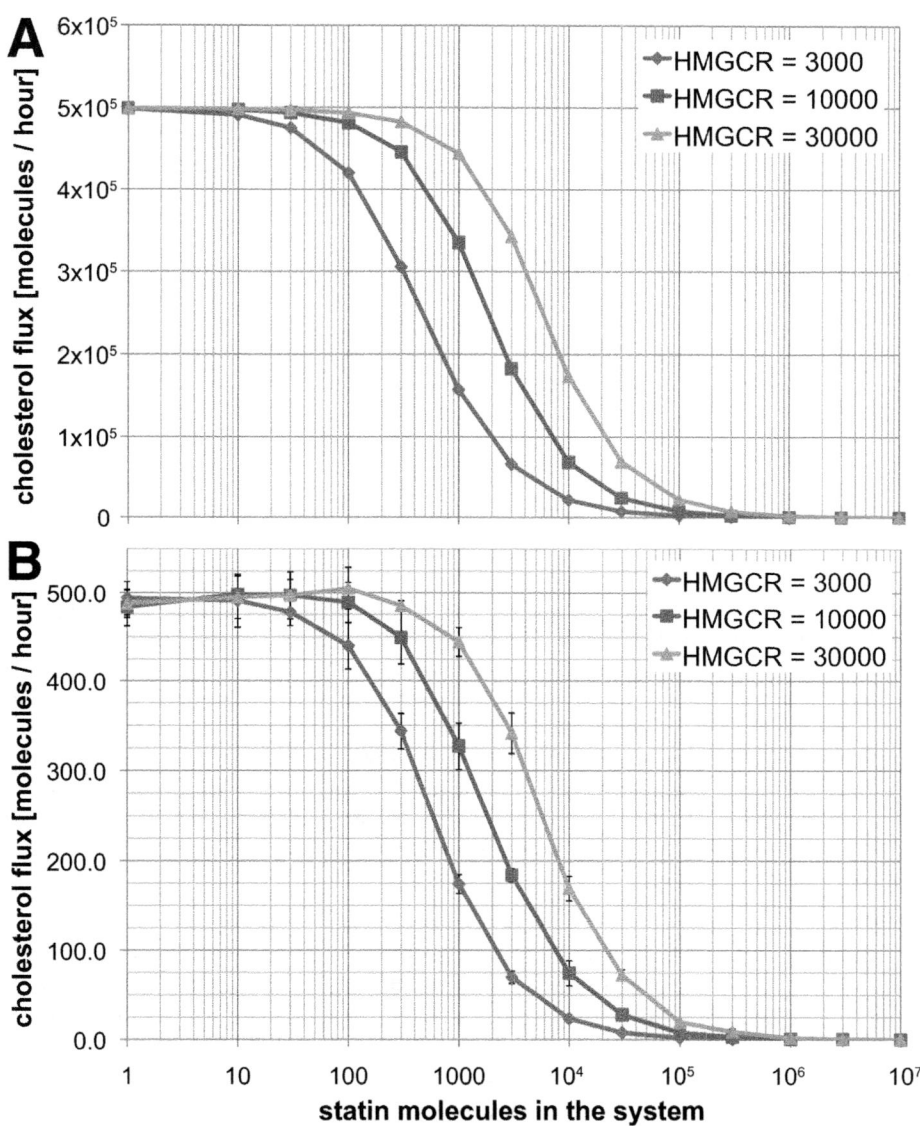

Fig. 5. Response of cholesterol flux to different amounts of statin molecules in the system. (A) Assuming a high influx of acetyl-CoA as computed by the Adaptive Dormant Prince ODE solver predicts mean expected values. This approximation works well for large molecule counts. (B) Assuming a low influx of acetyl-CoA as computed by the Gibson-Bruck stochastic simulator shows the variability associated with low copy numbers of molecules. Values were computed by the the Bio-PEPA Plugin 0.1.7 and report the average of five runs with error bars denoting standard deviations. We deliberately avoided averaging over many more repeats to highlight the stochastic nature of the system.

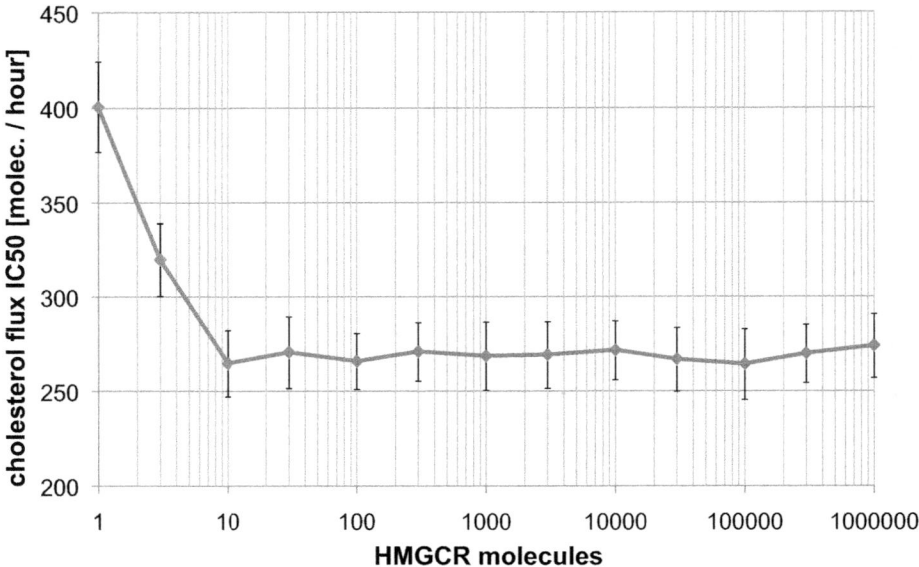

Fig. 6. The stochastic variability of the flux of cholesterol for a wide range of enzyme copy numbers with a corresponding number of inhibitory statin molecules as given below. The higher flux for 1 and 3 HMGCR molecules is caused by rounding off fractions computed by equation (4) to get molecule counts. Error bars denote standard deviations observed in 50 stochastic runs, measuring flux in the last hour of 2h simulations, as computed by the Gibson-Bruck stochastic simulator. The number of statin molecules used in the simulations shown here for a given HMGCR count were computed by rounding the result of equation (4). This resulted in the following HMGCR → statin pairs: $1 \to 0$; $3 \to 1$; $10 \to 6$; $30 \to 18$; $100 \to 62$; $300 \to 188$; $1000 \to 628$; $3000 \to 1885$; $1 \times 10^4 \to 6285$; $3 \times 10^4 \to 18857$; $1 \times 10^5 \to 62857$; $3 \times 10^5 \to 188571$; $1 \times 10^6 \to 628571$.

Visual inspection of all time courses with only 1 statin molecule present indicates that the cholesterol increase follows a straight line almost immediately from the start. This is confirmed by comparisons between the fluxes measured over the first and last quarter of the first hour simulated in ODEs which differ by less than 5%, a difference that is exceeded by the stochastic noise in the low-flux system. For stochastic simulations, measuring flux over a whole hour integrates more events leading to less variance than measuring shorter intervals. Thus we report as F the number of cholesterol molecules synthesised in the first hour after starting the simulation with all metabolites at zero. Since this number varies in stochastic simulations, we report the mean and standard deviation of five simulations in this case. As discussed below, the structure of this system is such that eventually cholesterol production will always reach a level that is equivalent to the influx of acetyl-CoA, even though this may be unrealistic in a cell because HMG-CoA is degraded in some other way or produced at lower rates. Thus it is desirable to measure flux as early as possible in this system; hence we limited most of our measurements to the first hour.

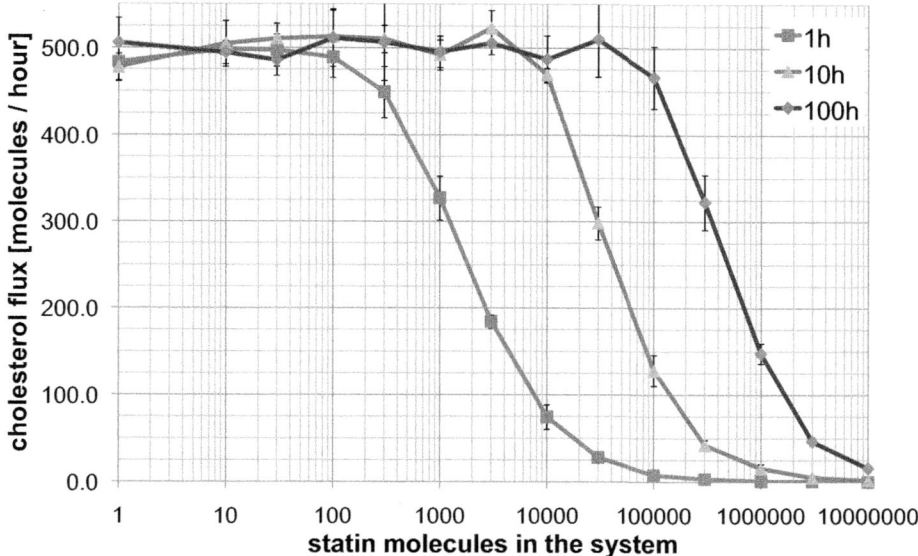

Fig. 7. Statin loses its inhibitory power if enough HMG-CoA accumulates over time in the absence of other degradation routes. The error bars denote the standard deviation of five different stochastic runs, as computed by the Gibson-Bruck stochastic simulator.

6.2   Simulation Results and Biological Interpretation

Statins (or HMG-CoA reductase inhibitors) are drugs widely-used to lower cholesterol levels [8,31,2]. They act by inhibiting the production of mevalonate that is catalysed by HMGCR, a step widely believed to be rate limiting for cholesterol production [21]. Much previous work has investigated this step in isolation [21,30,37], but little is known about the quantitative dynamics of the whole pathway. Here we analyse a model that provides the opportunity to quantitatively investigate the dynamics of the whole pathway. We have chosen values for unknown parameters that reflect the intuition of many biologists that HMGCR is rate limiting. This provides an optimal starting point for exploring the potential of statin to inhibit cholesterol production.

More specifically we are interested in evaluating how the function of statin is affected by natural diversity in the rate at which HMGCR catalyses the reaction that is blocked competitively by statin [30,37]. Such diversity in rate can come from variation in the numbers of enzymes per cell (e.g. different transcription, translation and degradation rates) or from variation in the turnover of the enzyme as caused by point mutations affecting its catalytic centre. We simulated the model in two settings, one with a high flux of acetyl-CoA using ODEs and one with a low flux of acetyl-CoA using stochastic simulations, reflecting conditions when the cell directs acetyl-CoA elsewhere. For each set we chose effective numbers of HMGCR = 3000, 10000 and 30000 copies per cell to capture natural diversity. We then measured for each of these six sets (3x ODE, 3x stochastic simulations) the flux of cholesterol at 14 different statin molecule counts in the system, spanning over seven orders of magnitude.

The ODE analysis in Figure 5A shows that cells with higher effective concentrations of HMGCR require larger doses of statin to shut down cholesterol synthesis. Repeating the same for a low flux of acetyl-CoA using stochastic simulations confirms this and indicates that the flux of acetyl-CoA into this pathway does not affect the relative power of statins to shut down cholesterol production (although it does affect the absolute amount produced; see y-axes in Figure 5 for comparison). Changes appear to be linear in that 10x more HMGCR requires 10x more statin to block and a 1000x higher flux requires 1000x more statin to reduce it to the level of the low flux we observed. The slight increase in flux with some of the numbers of statin molecules in Figure 5B is not significant (see error bars).

To investigate the effects of statin on the variability of flux at very low copy numbers of HMGCR we calculated the analytically expected number of statin molecules that blocks 50% of the flux of acetyl-CoA to cholesterol under a regime that leads to a local equilibrium of 500 molecules of HMG-CoA (this is similar to our low-flux regime). The expected number of inhibitor molecules I that achieves this effect is given by

$$I = K_I \Omega \frac{k_{cat} ES - STF - TFK_M \Omega}{TFK_M \Omega} \tag{4}$$

where E counts the enzyme HMGCR, S counts the substrate HMG-CoA (assumed to be 500), $k_{cat} = 500$ [1/h], $K_M = 0.07$ [mM], $\Omega = 1000$, T is the target flux to cholesterol assumed to be 500 [1/h], $F = 50\%$ is the fraction to which the flux should be reduced by statin and $K_I = 0.000044$ [mM] is the inhibition constant of statin. Figure 6 shows that the stochastic variability of the flux of cholesterol does not depend on the enzyme copy number although it is not possible to adjust the flux precisely for very few enzymes since the number of statin molecules has to be an integer (rounding off caused the higher flux in Figure 6).

Our model also allows us to investigate acquired tolerance towards statin as caused by the structure of the pathway. Figure 7 shows a comparison of the low flux environment with HMGCR = 10000, as observed over 1 h, 10 h and 100 h (each measured flux averages only over the last hour before the end of the observation interval, where the observations start in equilibrium at 0 h with the addition of statin and end after the specified time). Assuming that HMG-CoA is not degraded by alternative pathways and all reactions are irreversible, a more than 10x higher statin concentration is needed to block cholesterol production over 10 h than when only 1 h needs to be blocked.

Shutting down cholesterol production by competitive inhibition in our model leads to a continuous buildup of HMG-CoA since this metabolite is continuously produced and is not otherwise degraded. Because inhibition depends on an excess of statin in comparison to the metabolite HMG-CoA (see equation 2), given enough time the buildup of HMG-CoA will overpower any number of inhibitor molecules, making the pathway tolerant to the number of inhibiting molecules applied. This is demonstrated by the need for higher statin molecule numbers to shut down cholesterol production over longer periods of time (see Figure 7). In real cells an unbounded increase of any metabolite is not possible and might even be actively avoided by cells, thus acquired statin tolerance is limited in a natural setting. Nevertheless these findings indicate that the flexibility of pathways in circumventing obstacles needs to be considered in addition to variability in HMGCR levels and acetyl-CoA flux when calculating the right dose of statins.

7 Related Work

There are various languages associated with tools that map visual diagrams to quantitative modelling environments (e.g. SPiM [40], BlenX [14], Kappa [12], Snoopy [22], EPN-PEPA [45], JDesigner [44]). However the corresponding graphical notations are not as rich as SBGN-PD and are thus not easily applied to the wide range of scenarios that SBGN-PD was designed for. Since SBGN-PD is emerging as a new standard, it is clearly desirable to translate from SBGN-PD to a quantitative environment.

Since the first draft of SBGN-PD has been published in August 2008, a number of tools have been developed to support it, including the Edinburgh Pathway Editor [46], Arcadia for visualisation [47] , TinkerCell that is linked to the Systems Biology Workbench [6], and PathwayLab [26]. The graphical editor CellDesigner [18] supports a subset of SBGN-PD and can translate it into SBML which is supported by many quantitative analysis tools. However the process of adding quantitative information involves cumbersome manual interventions. This motivated work for SBMLsqueezer [16], a CellDesigner plug-in that supports the automatic construction of generalised mass action kinetics equations. While the automated suggestions for the kinetic laws from SBMLsqueezer might be of interest for some problems, the generated reactions contain many parameters that are extraordinarily difficult to estimate. Thus it is preferable to also allow the user to enter arbitrary kinetic laws that may have to be hand-crafted, but whose equations are simpler and require fewer parameter estimates. In SBGNtext2BioPEPA this is combined with mechanisms to reuse the code for such kinetic laws, greatly reducing practical difficulties and the potential for errors.

8 Conclusion and Perspectives

Since biologists are much more comfortable with drawing visual diagrams than writing code, support for translating SBGN-PD into quantitative analysis frameworks can play a key role in facilitating quantitative modelling. Our experiences with modelling the cholesterol pathway have highlighted the value of quick access to details of the model from an SBGN-PD compliant editor like Edinburgh Pathway Editor. The tool-chain described here efficiently transforms a graphical SBGN-PD model into SBGNtext, which is then compiled into a Bio-PEPA model that is ready for simulation. The simulation results presented here show that this system can indeed be used for analysing non-trivial questions.

The workflow presented here critically depends on the process flow abstraction that implicitly underlies SBGN-PD. We have explicitly described this process flow abstraction and used it to design a mechanism for translating SBGN-PD into a computational model that can be used for quantitative analysis. In order to do this we build on SBGNtext, a textual representation of SBGN-PD that we created [34,35] and that focusses on the key functional SBGN-PD content, avoiding the clutter that comes from storing graphical details. We have developed our translator SBGNtext2BioPEPA in Java to facilitate its integration with the Bio-PEPA Eclipse Plugin and the Systems Biology Software Infrastructure (SBSI) that is currently under development at the Centre for Systems Biology at Edinburgh (http://csbe.bio.ed.ac.uk/). SBGNtext2BioPEPA contains a parser for SBGNtext based on a formal ANTLR EBNF grammar and is freely available [34]. Building on the process flow abstraction and the internal representation of entities, processes, arcs and parameters in our code facilitates implementing translations of SBGNtext to other modelling languages.

Acknowledgements. We thank Stephen Gilmore for helpful comments that improved this manuscript. The Centre for Systems Biology Edinburgh is a Centre for Integrative Systems Biology (CISB) funded by BBSRC and EPSRC, reference BB/D019621/1.

References

1. Akman, O.E., Guerriero, M.L., Loewe, L., Troein, C.: Complementary approaches to understanding the plant circadian clock. In: Proc. of FBTC 2010. EPTCS, vol. 19, pp. 1–19 (2010)
2. Baigent, C., Keech, A., Kearney, P.M., Blackwell, L., Buck, G., Pollicino, C., Kirby, A., Sourjina, T., Peto, R., Collins, R., Simes, R.: Efficacy and safety of cholesterol-lowering treatment: prospective meta-analysis of data from 90,056 participants in 14 randomised trials of statins. Lancet 366, 1267–1278 (2005)
3. Bio-PEPA homepage, http://www.biopepa.org/; To install the Bio-PEPA Eclipse Plugin by Adam Duguid follow the links from
   http://homepages.inf.ed.ac.uk/jeh/Bio-PEPA/Tools.html (2009)
4. Calder, M., Duguid, A., Gilmore, S., Hillston, J.: Stronger computational modelling of signalling pathways using both continuous and discrete-state methods. In: Priami, C. (ed.) CMSB 2006. LNCS (LNBI), vol. 4210, pp. 63–77. Springer, Heidelberg (2006)
5. Calder, M., Hillston, J.: Process algebra modelling styles for biomolecular processes. In: Priami, C., Back, R.-J., Petre, I. (eds.) Transactions on Computational Systems Biology XI. LNCS (LNBI), vol. 5750, pp. 1–25. Springer, Heidelberg (2009)
6. Chandran, D., Bergmann, F., Sauro, H.: TinkerCell: modular CAD tool for synthetic biology. Journal of Biological Engineering 3(1), 19 (2009), http://www.tinkercell.com
7. Chang, A., Scheer, M., Grote, A., Schomburg, I., Schomburg, D.: BRENDA, AMENDA and FRENDA the enzyme information system: new content and tools in 2009. Nucleic Acids Res. 37, D588–D592 (2009), http://www.brenda-enzymes.org/
8. Chasman, D.I., Posada, D., Subrahmanyan, L., Cook, N.R., Stanton Jr., V.P., Ridker, P.M.: Pharmacogenetic study of statin therapy and cholesterol reduction. JAMA 291, 2821–2827 (2004)
9. Ciocchetta, F., Gilmore, S., Guerriero, M.L., Hillston, J.: Integrated Simulation and Model-Checking for the Analysis of Biochemical Systems. In: Proc. of PASM 2008. ENTCS, vol. 232, pp. 17–38 (2009)
10. Ciocchetta, F., Hillston, J.: Bio-PEPA: a Framework for the Modelling and Analysis of Biological Systems. Theoretical Computer Science 410(33-34), 3065–3084 (2009)
11. Cytoscape Consortium: Cytoscape Home page (2009), http://cytoscape.org/
12. Danos, V., Feret, J., Fontana, W., Harmer, R., Krivine, J.: Rule-based modelling of cellular signalling. In: Caires, L., Li, L. (eds.) CONCUR 2007. LNCS, vol. 4703, pp. 17–41. Springer, Heidelberg (2007)
13. Danos, V., Laneve, C.: Formal molecular biology. Theoretical Computer Science 325, 69–110 (2004)

14. Dematté, L., Priami, C., Romanel, A.: The BlenX Language: A Tutorial. In: Bernardo, M., Degano, P., Tennenholtz, M. (eds.) SFM 2008. LNCS, vol. 5016, pp. 313–365. Springer, Heidelberg (2008)
15. Demir, E., Babur, O., Dogrusoz, U., Gursoy, A., Nisanci, G., Cetin-Atalay, R., Ozturk, M.: PATIKA: an integrated visual environment for collaborative construction and analysis of cellular pathways. Bioinformatics 18, 996–1003 (2002)
16. Draerger, A., Hassis, N., Supper, J., Schröder, A.Z.: SBMLsqueezer: A CellDesigner plug-in to generate kinetic rate equations for biochemical networks. BMC Systems Biology 2, 39 (2008)
17. Duguid, A., Gilmore, S., Guerriero, M.L., Hillston, J., Loewe, L.: Design and Development of Software Tools for Bio-PEPA. In: Proc. of WSC 2009, pp. 956–967. IEEE Press, Los Alamitos (2009)
18. Funahashi, A., Matsuoka, Y., Jouraku, A., Morohashi, M., Kikuchi, N., Kitano, H.: CellDesigner 3.5: A Versatile Modeling Tool for Biochemical Networks. Proceedings of the IEEE 96(issue 8), 1254–1265 (2008), http://www.celldesigner.org/
19. Gibson, M.A., Bruck, J.: Efficient Exact Stochastic Simulation of Chemical Systems with Many Species and Many Channels. J. Phys. Chem. 104, 1876–1889 (2000)
20. Gillespie, D.T.: Stochastic Simulation of Chemical Kinetics. Annu. Rev. Phys. Chem. 58, 35–55 (2007)
21. Goldstein, J.L., Brown, M.S.: Regulation of the mevalonate pathway. Nature 343, 425–430 (1990)
22. Heiner, M., Richter, R., Schwarick, M., Rohr, C.: Snoopy – A tool to design and execute graph-based formalisms. Petri Net Newsletter 74, 8–22 (2008), http://www-dssz.informatik.tu-cottbus.de/software/snoopy.html
23. Hillston, J.: A Compositional Approach to Performance Modelling. Cambridge University Press, Cambridge (1996)
24. Hlavacek, W.S., Faeder, J.R., Blinov, M.L., Posner, R.G., Hucka, M., Fontana, W.: Rules for modeling signal-transduction systems. Science STKE 344, re6 (2006)
25. Hucka, M., Hoops, S., Keating, S., Le Novère, N., Sahle, S., Wilkinson, D.: Systems Biology Markup Language (SBML) Level 2 Version 4 Release 1. Nature Proceedings (2008), http://dx.doi.org/10.1038/npre.2008.2715.1 and http://sbml.org/Documents/Specifications
26. Jansson, A., Jirstrand, M.: Biochemical modeling with Systems Biology Graphical Notation. Drug Discovery Today (2010)
27. Kanehisa, M., Goto, S., Furumichi, M., Tanabe, M., Hirakawa, M.: KEGG for representation and analysis of molecular networks involving diseases and drugs. Nucleic Acids Res. 38, D355–D360 (2010), http://www.genome.jp/kegg/
28. Kitano, H., Funahashi, A., Matsuoka, Y., Oda, K.: Using process diagrams for the graphical representation of biological networks. Nature Biotechnology 23, 961–966 (2005)
29. Kohn, K.W., Aladjem, M.I., Kim, S., Weinstein, J.N., Pommier, Y.: Depicting combinatorial complexity with the molecular interaction map notation. Mol. Syst. Biol. 2, 51 (2006)
30. Krauss, R.M., Mangravite, L.M., Smith, J.D., Medina, M.W., Wang, D., Guo, X., Rieder, M.J., Simon, J.A., Hulley, S.B., Waters, D., Saad, M., Williams, P.T., Taylor, K.D., Yang, H., Nickerson, D.A., Rotter, J.I.: Variation in the 3-hydroxyl-3-methylglutaryl coenzyme a reductase gene is associated with racial differences in low-density lipoprotein cholesterol response to simvastatin treatment. Circulation 117, 1537–1544 (2008)
31. Law, M.R., Wald, N.J., Rudnicka, A.R.: Quantifying effect of statins on low density lipoprotein cholesterol, ischaemic heart disease, and stroke: systematic review and meta-analysis. BMJ 326, 1423 (2003)

32. Le Novère, N., Hucka, M., Mi, H., Moodie, S., Schreiber, F., Sorokin, A., Demir, E., Wegner, K., Aladjem, M.I., Wimalaratne, S.M., Bergman, F.T., Gauges, R., Ghazal, P., Kawaji, H., Li, L., Matsuoka, Y., Villeger, A., Boyd, S.E., Calzone, L., Courtot, M., Dogrusoz, U., Freeman, T.C., Funahashi, A., Ghosh, S., Jouraku, A., Kim, S., Kolpakov, F., Luna, A., Sahle, S., Schmidt, E., Watterson, S., Wu, G., Goryanin, I., Kell, D.B., Sander, C., Sauro, H., Snoep, J.L., Kohn, K., Kitano, H.: The Systems Biology Graphical Notation. Nature Biotechnology 27, 735–741 (2009)
33. Le Novère, N., Moodie, S., Sorokin, A., Hucka, M., Schreiber, F., Demir, E., Mi, H., Matsuoka, Y., Wegner, K., Kitano, H.: Systems Biology Graphical Notation: Process Diagram Level 1. Nature Preceedings (2008), http://hdl.handle.net/10101/npre.2008.2320.1
34. Loewe, L.: The SBGNtext2BioPEPA homepage (2009), http://csbe.bio.ed.ac.uk/SBGNtext2BioPEPA/index.php
35. Loewe, L., Moodie, S., Hillston, J.: Defining a textual representation for SBGN Process Diagrams and translating it to Bio-PEPA for quantitative analysis of the MAPK signal transduction cascade. Tech. rep., School of Informatics, University of Edinburgh (2009), http://csbe.bio.ed.ac.uk/SBGNtext2BioPEPA/index.php
36. Loewe, L., Moodie, S., Hillston, J.: Quantifying the implicit process flow abstraction in SBGN-PD diagrams with Bio-PEPA. In: Proc. of CompMod 2009. EPTCS, vol. 6, pp. 93–107 (2009), http://arxiv.org/abs/0910.1410
37. Medina, M.W., Gao, F., Ruan, W., Rotter, J.I., Krauss, R.M.: Alternative splicing of 3-hydroxy-3-methylglutaryl coenzyme A reductase is associated with plasma low-density lipoprotein cholesterol response to simvastatin. Circulation 118, 355–362 (2008)
38. Moodie, S.L., Sorokin, A., Goryanin, I., Ghazal, P.: A Graphical Notation to Describe the Logical Interactions of Biological Pathways. J. Integr. Bioinformatics 3(2), 36 (2006)
39. Parr, T.: The Definitive ANTLR Reference: Building Domain-Specific Languages. The Pragmatic Bookshelf, Raleigh (2007), http://www.antlr.org/
40. Phillips, A.: A Visual Process Calculus for Biology. In: Symbolic Systems Biology: Theory and Methods, Jones and Bartlett Publishers (to appear, 2010), http://research.microsoft.com/en-us/projects/spim/
41. Priami, C.: Stochastic π-calculus. The Computer Journal 38(7), 578–589 (1995)
42. Ramsey, S., Orrell, D., Bolouri, H.: Dizzy: stochastic simulation of large-scale genetic regulatory networks. J. Bioinf. Comp. Biol. 3(2), 415–436 (2005), http://magnet.systemsbiology.net/software/Dizzy/
43. Raza, S., Robertson, K.A., Lacaze, P.A., Page, D., Enright, A.J., Ghazal, P., Freeman, T.C.: A logic-based diagram of signalling pathways central to macrophage activation. BMC Syst. Biol. 2, 36 (2008)
44. Sauro, H.M., Hucka, M., Finney, A., Wellock, C., Bolouri, H., Doyle, J., Kitano, H.: Next generation simulation tools: the Systems Biology Workbench and BioSPICE integration. OMICS 7(4), 355–372 (2003), For the graphical front end JDesigner http://www.sys-bio.org/software/jdesigner.htm
45. Shukla, A.: Mapping the Edinburgh Pathway Notation to the Performance Evaluation Process Algebra. Master's thesis, University of Trento, Italy (2007)
46. Sorokin, A., Paliy, K., Selkov, A., Demin, O., Dronov, S., Ghazal, P., Goryanin, I.: The Pathway Editor: A tool for managing complex biological networks. IBM J. Res. Dev. 50, 561–573 (2006), http://www.bioinformatics.ed.ac.uk/epe/; This work used the Edinburgh Pathway Editor prototype version EPE-3.0.0-alpha13 from http://epe.sourceforge.net/SourceForge/EPE.html
47. Villéger, A.C., Pettifer, S.R., Kell, D.B.: Arcadia: a visualization tool for metabolic pathways. Bioinformatics 26(11), 1470–1471 (2010)

Impulse-Based Dynamic Simulation of Deformable Biological Structures

Rhys Goldstein and Gabriel Wainer

Carleton University, Ottawa ON K1S5B6, Canada

Abstract. We present a new impulse-based method, called the Tethered Particle System (TPS), for the dynamic simulation of deformable biological structures. The TPS is unusual in that it may capture a gradual process of deformation using only instantaneous impulses that occur in response to particle collisions. This paper describes the method and its application to synaptic vesicle clusters and deformable biological membranes. Unlike many alternative methods, which require solutions to systems of equations or inequalities, the calculations in a TPS simulation are all analytic. The TPS also alleviates the need to choose regular time intervals appropriate for biological entities that may differ in size by orders of magnitude. The method is promising for simulations of small-scale self-assembling deformable biological structures exhibiting random motion.

1 Introduction

Simulation is becoming an increasingly common tool among biologists and medical researchers, complementing traditional experimental techniques. As Kitano explains in [11], experimental data is first used to form a hypothesis, and that hypothesis may be investigated with a simulation. Predictions made by the simulation can then be tested using *in vitro* and *in vivo* studies, and the new experimental data may lead to new hypotheses. This iterative process can be applied to basic research on biological systems, as well the development of drugs and other treatments.

Modeling and simulation methods that capture the dynamics of deformable biological structures are frequently targeted at surgical planning and training [3], as well as the analysis of prosthetics [8]. Models of smaller-scale deformable biological structures are rarer, but examples include the simulated deformation of 8-μm red blood cells [19], and that of membrane-sculpting proteins on the 10-nm scale [12].

The most common methods for simulating the dynamics of deformable structures are mass-spring-damper systems and the finite element method [7]. Our method, the Tethered Particle System (TPS), differs in that it uses only impulses to alter motion. Impulse-based methods have previously been used to simulate rigid bodies, but are generally neglected or considered unsuitable for objects that deform. It is counterintuitive to model deformable structures with impulses, as

C. Priami et al. (Eds.): Trans. on Comput. Syst. Biol. XIII, LNBI 6575, pp. 39–60, 2011.

impulses are instantaneous whereas the deformation of an object is a continuous process that may require a significant length of time. Nevertheless, if one represents a deformable structure as a network of a large number of particles, then numerous collisions and impulses between those particles may produce the effect of a gradual deformation of the overall structure.

We demonstrate that impulse-based methods provide a relatively simple way to allow deformable biological structures to assemble themselves from rigid particles representing proteins and other biological entities. Also, with an impulse-based simulation, it is easy to incorporate the random motion exhibited by these small biological objects. The TPS proved useful for the simulation of deformable vesicle clusters in a presynaptic nerve terminal, which form from interactions between randomly-moving synaptic vesicles and synapsin protein.

Section 2 provides an overview of existing dynamic simulation methods for both rigid bodies and deformable structures. Section 3 describes the TPS method in detail, and Section 4 presents TPS models of synaptic vesicle clusters and various deformable membranes. Strengths and weaknesses of the new method are discussed in Section 5.

2   Dynamic Simulation Methods

We use the phrase "dynamic simulation" to indicate the simulation of motion using laws of classical dynamics. Here we describe several pre-existing dynamic simulation methods, some intended for rigid bodies and others designed for deformable structures.

2.1   Dynamic Simulation of Rigid Bodies

This section reviews methods for the dynamic simulation of rigid bodies. In these methods, object deformation may be represented by a loss of kinetic energy, for example, or an overlapping of objects. If an object's changing shape is modeled, however, then we classify it as a deformable structure instead of a rigid body.

"Impulse-based" methods are perhaps the most obvious approach to the dynamic simulation of rigid bodies. An impulse-based method involves two tasks: collision detection (the task of calculating the time at which any two objects come into contact) and collision response (the task of computing the new trajectories of two colliding objects). In response to a collision, the trajectory of an object changes instantaneously in simulated time. The instantaneous change in the momentum of the object is referred to as an "impulse" [13].

Because impulse-based methods assume instantaneous contacts, the approach seems inappropriate for the modeling of stable contacts. If a ball is rolling across a table, for example, it remains in contact with the table for a length of time. In his 1996 Ph.D. thesis, Brian Mirtich demonstrated that stable contacts could be modeled as sequences of independent collisions [14]. Consider an impulse-based simulation of an object bouncing along a horizontal surface. Provided each bounce was sufficiently short in height and duration, the model could accurately represent a ball rolling across a table.

Another well-known drawback to impulse-based methods is possibility of simultaneous or nearly-simultaneous collisions. Consider a situation in which a small object is directly between two much larger approaching objects. After the first large object hits it, the small object may end up travelling back and forth between the larger objects in a long sequence of nearly-simultaneous collisions. This might require considerable computational effort. If kinetic energy is lost in each collision, then it is possible for the sequence of collisions to become infinite, slowing the simulation to a halt.

One advantage to impulse-based simulation is the simplicity of the method. If one is to implement an algorithm to detect collisions between pairs of objects, which is necessary in perhaps all of the competing methods, it is a simple matter to apply the law of conservation of momentum to give the two objects new trajectories. The constraint-based method of [1], by contrast, may compute new trajectories for more than two objects simultaneously. This is done by minimizing a linear function constrained by a system of linear inequalities. Depending on the model, this problem may be NP-hard, may have no solutions, or may have multiple different solutions.

Both impulse-based methods and the constraint-based method of [1] prevent the penetration of objects. "Penalty methods" differ in that they allow approaching objects to overlap slightly upon colliding. Typically, a spring is temporarily inserted between colliding objects; the more the objects overlap, the stronger the restoring force of the compressed spring [15].

2.2 Dynamic Simulation of Deformable Structures

We now review methods for the dynamic simulation of deformable structures, as opposed to rigid bodies. Note that phrase "dynamic simulation" excludes a wide range of methods for modeling deformable structures. An algorithm that fits a spline to a cross-sectional image of a human lung, for example, certainly models a deformable structure. But unless the motion of the lung is predicted from laws of physics, we would not consider it a dynamic simulation.

One way to model a deformable structure is with a set of point masses. Each mass is connected to its neighbors with a spring and possibly a damper. A spring applies a force that, depending on its present length, attracts or repels the masses on either end. A damper applies a force that decreases the relative speed of the masses on either end. These "mass-spring-damper" systems can be used to simulate the dynamics of deformable structures by predicting the acceleration of each mass, at regular time intervals, according to spring, damper, and external forces [16]. The mass-spring-damper method is essentially a penalty method like those described in Section 2.1 for rigid bodies. The difference is that the springs in Section 2.1 are inserted temporarily between detached colliding objects, whereas in this case the springs tend to be permanent and the point masses do not necessarily represent distinct objects.

Some mass-spring-damper models use spherical particles instead of point masses. This technique is used in [5] to simplify the detection of collisions between deformable objects. Each object is composed of several overlapping spherical particles. Instead of detecting collisions between the possibly-concave surfaces of these objects, only collisions between particles are considered. A similar approach is taken in [9], which incorporates friction, viscous forces, and the fracture of deformable objects.

Mass-spring-damper methods are used extensively in computer graphics. They are considered computationally efficient, but not particularly accurate. Incompressible deformable objects and nearly-rigid thin membranes are difficult to model, and appropriate spring parameters may be difficult to determine. Stiff objects, modeled using springs with large restoring forces, threaten the stability of mass-spring-damper simulations. Techniques have been developed to address the stiffness problem. The simplest solution is to decrease the time step, though this increases computational costs.

A popular alternative to mass-spring-damper systems is the "finite element method" (FEM) [7]. FEM actually refers to a more general mathematical technique, but we will refer to it as a dynamic simulation method for deformable structures. In an FEM model, a deformable object is represented as a set of adjacent polyhedra. Each polyhedron, or "element", has a set of vertices, or "nodes". Although recorded attributes are associated with each node, material properties can be obtained at every point in each element by interpolating the attributes of each node. Positions of each node may change at each time step. FEM simulations are considered to be more accurate, but also more computationally intensive, than those based on mass-spring-damper models. The FEM is most efficient with metals and materials that exhibit relatively little deformation. Highly deformable materials, like soft biological tissues, require frequent re-calculation of large mass and stiffness matrices that depend on the positions of the nodes.

Impulse-based methods, like those used for rigid bodies, tend to be either neglected or avoided for the dynamic simulation of deformable structures. A literature search revealed an "impulse response deformation model" [18], which does simulate the dynamics of deformable structures, but is not an impulse-based method despite its name. In this case the term "impulse" refers to an initial perturbation in an object's shape. Convolution integrals are used to track the object's shape after the perturbation.

The possibility of applying impulse-based methods to deformable objects is acknowledged in [10], but quickly dismissed with the assertion that "impulse-based methods assume short contacts only, and therefore they are not suitable for soft objects". The argument is intuitive: impulses are instantaneous changes in momentum, whereas the deformation of an object is a gradual process that takes place over time. In [14], Mirtich states that the strongest restriction of impulse-based methods is that models are comprised of only rigid bodies.

The pre-existing method that most closely fits the phrase "impulse-based dynamic simulation of deformable structures" was developed recently to simulate inextensible cloth [2], as well as volume-conserving deformable objects [6]. In both cases, impulses are applied simultaneously to all particles in a structure at regular intervals. Because the purpose of these impulses is to constrain either the distances or volumes between the particles, the method can be classified as constraint-based as well as impulse-based. The simulations of [2] and [6] differ from impulse-based rigid body simulations in that, in the latter, impulses occur in response to collisions and not at regular intervals.

3   Tethered Particle System Method

A TPS model tracks the positions and velocities of numerous particles, each with a fixed mass, that interact with one another via collisions. The TPS method is unusual in that it is an impulse-based method, meaning that any change in a particle's velocity is instantaneous, yet the method is designed for representing structures that deform over a length of time. The key idea is that a deformable structure may be represented by a group of particles; even though each individual particle changes velocity in a sequence of instantaneous impulses, the configuration of the particles in the group changes in a seemingly gradual process over time. This section provides a detailed description of the TPS method, including key equations.

3.1   Blocking and Tethering Collisions

In order for a group of particles to exhibit any structure at all, the distances between certain pairs of particles in the group must be regulated or restricted. In a TPS model, the distance between a pair of particles is constrained by two types of collisions: "blocking collisions" and "tethering collisions".

What we refer to as a "blocking collision" is what one normally associates with the word "collision". As illustrated in Figure 1a, a blocking collision occurs when two approaching particles reach an inner limiting distance. This "blocking distance" is represented by $\Delta u_{blocking}$.

Note that although we will frequently depict a particle as a circle or sphere of some radius, neither the shape nor the size of a particle is explicitly defined in a TPS model. The particles in Figure 1 could have been drawn as larger circles, for instance, perhaps overlapping with one another at the time of collision.

The particles in Figure 1a are shown rebounding at a similar angle to that at which they had been approaching. This indicates that the collision is elastic, meaning that no kinetic energy is lost. When real-world objects collide, they deform and absorb kinetic energy. Although individual particles in a TPS have no shape and do not explicitly deform, we may wish to account for energy loss in particle collisions. The loss of kinetic energy in the inelastic collision of Figure 1b causes the particles to rebound at a smaller angle.

Energy loss due to collisions is generally modeled with a parameter called the "coefficient of restitution", which expresses the ratio of the post-collision relative speed of two particles to the pre-collision relative speed [1]. In a TPS model, different coefficients are used for different types of collisions. In a blocking collision, the coefficient of restitution is referred to as the "rebounding coefficient", and is represented by $c_{rebound}$. We will assume $0 \leq c_{rebound} \leq 1$, with $c_{rebound} = 0$

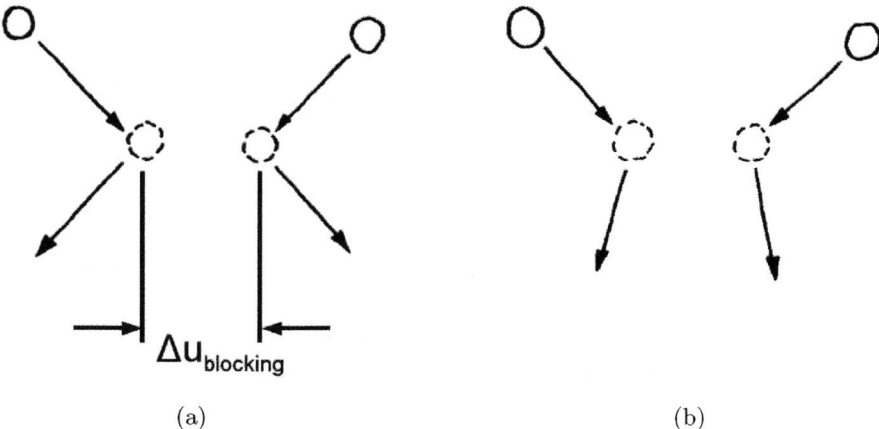

Fig. 1. Illustrations of blocking collisions, in which pairs of approaching particles reach the blocking distance $\Delta u_{blocking}$ and rebound. In (a) the collision is elastic, whereas (b) depicts an inelastic collision.

indicating maximum energy loss, and $c_{rebound} = 1$ indicating a perfectly elastic collision.

Any two specific particles may be tethered together at the start of a simulation. Also, when a blocking collision occurs between two particles, they may become tethered. If two particles are tethered and moving away from one another, and if they reach the "tethering distance" $\Delta u_{tethering}$, then one of two things may happen. The particles may become untethered, in which case they continue moving away from one another. Otherwise the particles remain tethered, undergo a tethering collision, and retract inwards. The phrase "tethering collision" is unintuitive because one normally expects a collision to occur only between approaching objects. We use the word "collision" for separating particles as well so that we can apply the phrases "collision detection" and "collision response" to either type of particle-particle interaction.

A tethering collision may be envisioned as a situation in which a cord has completely unravelled, and therefore delivers an inward impulse to the particles attached to it on either end. Such a cord is illustrated in Figure 2a. Initially, the cord is slack (solid line). As the particles move apart, the cord unravels and eventually becomes taut (dotted lines). At that point, the particles change direction and move inwards. Unlike a spring in a mass-spring-damper model, a cord in a TPS model has no effect on either particle before reaching its maximum length.

Figure 2a is meant to portray an elastic tethering collision, as the particles end up approaching one another at the same relative angle at which that had been separating. Figure 2b, by contrast, illustrates an inelastic tethering collision in which energy is lost. The particles end up approaching at a smaller angle. To address energy loss in tethering collisions, we introduce another type of coefficient of restitution. We refer to it as the "retraction coefficient", and represent it with $c_{retract}$ satisfying $0 \leq c_{retract} \leq 1$.

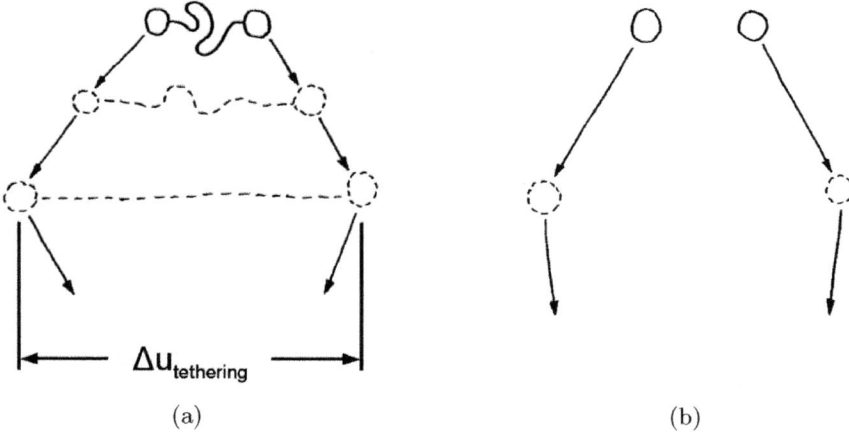

Fig. 2. Illustrations of tethering collisions, in which pairs of separating tethered particles reach the tethering distance $\Delta u_{tethering}$ and retract. In (a) the tethering collision is elastic, whereas (b) depicts an inelastic collision.

3.2   Basic Simulation Procedure

Suppose we have two particles, A and B. Their masses are m_A and m_B respectively, their current positions are u_A and u_B, and their velocities are v_A and v_B. The distance Δu between the particles can be expressed as a function of the time Δt.

$$\Delta u = \sqrt{\sum \left(((u_B + v_B \cdot \Delta t) - (u_A + v_A \cdot \Delta t))^2 \right)} \tag{1}$$

The basic procedure in a TPS simulation is to repeatedly solve (1) for Δt for all pairs of relatively close particles. Solving (1) with $\Delta u = \Delta u_{blocking}$ yields the time remaining before a blocking collision, whereas $\Delta u = \Delta u_{tethering}$ gives the time of a tethering collision. Time is advanced by the smallest calculated value of Δt, the time remaining before the next collision. When that collision occurs, the new velocities of the two particles involved are calculated from (2), and the process repeats.

$$
\begin{aligned}
v_A{}' &= v_A + \frac{\Delta p}{m_A} \\
v_B{}' &= v_B - \frac{\Delta p}{m_B}
\end{aligned}
\tag{2}
$$

The vector Δp above is the impulse, the change in momentum of particle A as a result of the collision. To obtain its value, it is useful to calculate the following vectors. Note that $v_{\hat{u}}$ is the relative velocity of the particles projected onto the axis between them.

$$\hat{u} = \frac{\hat{u}_B - \hat{u}_A}{\sqrt{\sum (\hat{u}_B - \hat{u}_A)^2}}$$
$$v_{AB} = v_B - v_A$$
$$v_{\hat{u}} = \sum (v_{AB} \cdot \hat{u}) \cdot \hat{u}$$

If the particles rebound or retract, then Δp can be calculated from (3) below with $c_{restitute}$ being either $c_{rebound}$ or $c_{retract}$.

$$\Delta p = \left(\frac{1}{m_A} + \frac{1}{m_B} \right)^{-1} \cdot (1 + c_{restitute}) \cdot v_{\hat{u}} \qquad (3)$$

The actual computations performed are complicated by the possibility of revolution, simultaneous and nearly-simultaneous collisions, and random impulses. These concepts are described in Sections 3.3, 3.4, and 3.6. The TPS remains simpler than most deformable structure simulation methods in that all unknown variables can be calculated analytically from explicit formulas. There are no systems of equations or inequalities that need to be solved simultaneously or iteratively.

3.3   Revolution

Note that it is the tethering collisions that distinguish the TPS from more traditional particle collision algorithms. They place potentially-useful outer limits on the distances between certain pairs of particles, but introduce a performance problem that must be addressed. The problem is illustrated in Figure 3. Particle A remains stationary because it has an infinite mass, whereas particle B is in motion with a finite mass. The two particles remain tethered, and undergo a sequence of elastic tethering collisions. Immediately after each collision, particle B approaches A at a relatively small angle θ.

Note that at any stage in a TPS simulation, time is advanced by an irregular interval to that of the next collision. The greater the frequency with which collisions occur, the slower the simulation progresses. The problem with the above scenario is that, if θ is small, the tethering collisions become extremely

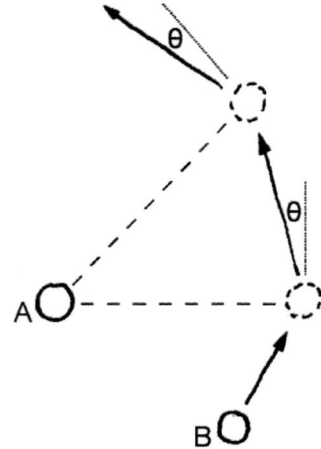

Fig. 3. A scenario in which particle B revolves around A in a sequence of tethering collisions

frequent and the simulation may become impractically slow. Worse, if $\theta = 0$, then time cannot be advanced at all without violating the constraint of $\Delta u_{tethering}$ on the distance between A and B. The problem still exists if the mass of A is finite, and is exacerbated by small values of $c_{retract}$.

To address the problem, we place a lower limit $\theta_{revolve}$ on the angle at which particles can approach after a tethering collision. In a collision where this restriction takes effect, we say that the particles "revolve" instead of "retract". We also introduce a "revolution coefficient" $c_{revolve}$ that expresses the ratio of the new relative velocity to the old one after one complete revolution of the particles, allowing energy to be lost. We require $0 \leq c_{revolve} \leq 1$.

Calculations pertaining to revolution require us to obtain $\boldsymbol{v}_{\hat{w}}$, the component of the relative velocity perpendicular to the axis between them.

$$\boldsymbol{v}_{\hat{w}} = \boldsymbol{v}_{AB} - \boldsymbol{v}_{\hat{u}}$$

If the condition in (4) is satisfied, then the retraction impulse calculated from (3) is sufficient.

$$\sqrt{\sum \left((c_{retract} \cdot \boldsymbol{v}_{\hat{u}})^2 \right)} > tan\,(\theta_{revolve}) \cdot \sqrt{\sum (\boldsymbol{v}_{\hat{w}}{}^2)} \tag{4}$$

If (4) is not satisfied, we abandon (3) and use the more general equation in (5).

$$\Delta \boldsymbol{p} = \left(\frac{1}{m_A} + \frac{1}{m_B} \right)^{-1} \cdot (\boldsymbol{v}_{AB} - \boldsymbol{v}_{AB}{}') \tag{5}$$

Here $\boldsymbol{v}_{AB}{}'$, the post-collision relative velocity, is obtained as follows.

$$\boldsymbol{v}_{revolve} = \boldsymbol{v}_{\hat{w}} - tan\,(\theta_{revolve}) \cdot \sqrt{\sum (\boldsymbol{v}_{\hat{w}}{}^2)} \cdot \hat{u}$$
$$\hat{u}_{revolve} = \frac{\boldsymbol{v}_{revolve}}{\sqrt{\sum (\boldsymbol{v}_{revolve})^2}}$$
$$\boldsymbol{v}_{AB}{}' = c_{revolve}{}^{\frac{\theta_{reolve}}{\pi}} \cdot \sqrt{\sum (\boldsymbol{v}_{AB}{}^2)} \cdot \hat{u}_{revolve}$$

If $\theta_{revolve} = \frac{\pi}{n}$, then it takes n collisions for the two particles to achieve a complete revolution, and the relative speed decreases by a factor of $c_{revolve}{}^{\frac{1}{n}}$ after each collision.

3.4   Loading and Restitution

It is widely known that simultaneous and nearly-simultaneous collisions threaten the efficiency of impulse-based methods, potentially slowing simulations to a halt. We now describe this problem followed by our solution.

Consider the scenario in Figure 4. Assume particles A, B, and C are all of the same mass, and that collisions are elastic. At time t_{AB}, moving particle A collides with stationary particle B, transferring all of its momentum. Then at time t_{BC}, particle B collides with stationary C and transfers all of its momentum.

An efficiency problem arises if the mass of particle B is reduced to a fraction of that of A and C, as illustrated in Figure 5. At time t_{AB_0}, particle A transfers

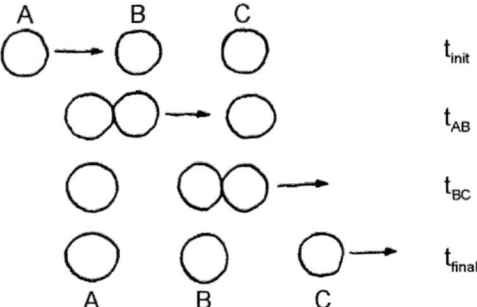

Fig. 4. A scenario in which 2 collisions occur between three particles of equal mass

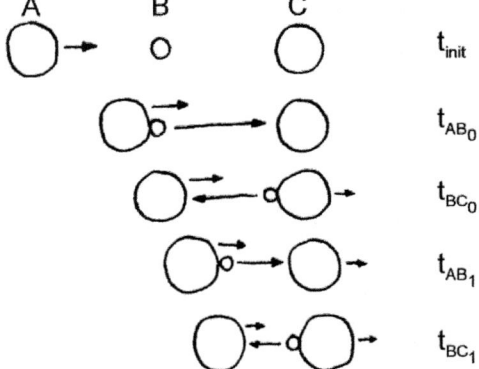

Fig. 5. The same scenario as in Figure 4, but the center particle is now less massive. Numerous nearly-simultaneous collisions result.

only some of its momentum to B. Particle B reaches C and rebounds at t_{BC_0}, then meets A again at time t_{AB_1}. Because only a small amount of momentum is transferred in each collision, particle B must rebound back and forth in a sequence of nearly-simultaneous collisions. If the mass of B is one thousandth that of A and C, roughly 70 elastic collisions occur before enough momentum has been transferred to separate all three particles.

The processing of 70 collisions is in itself a significant computational cost for such a simple scenario, but there are many situations in which the simulation will halt completely. When a simulation was performed with the Figure 5 scenario and a rebounding coefficient of 0.9, the momentum transferred on each collision eventually rounded to zero and the simulation stalled.

We propose a novel approximation that addresses the threat of simultaneous and nearly-simultaneous collisions. The idea is to separate each collision into a loading phase and a restitution phase, and to allow restitution to take place at a later time. When particles collide (the loading phase), they form loaded groups. A "loaded group" acts as a single body with the combined mass of all the particles in the group. A restitution delay time $\Delta t_{restitute}$ is introduced, after which the loaded particles separate (the restitution phase). Loaded particles may remain together longer than $\Delta t_{restitute}$ if necessary to ensure that the order in which particles separate is opposite that in which they loaded. Figure 6 illustrates the approximation.

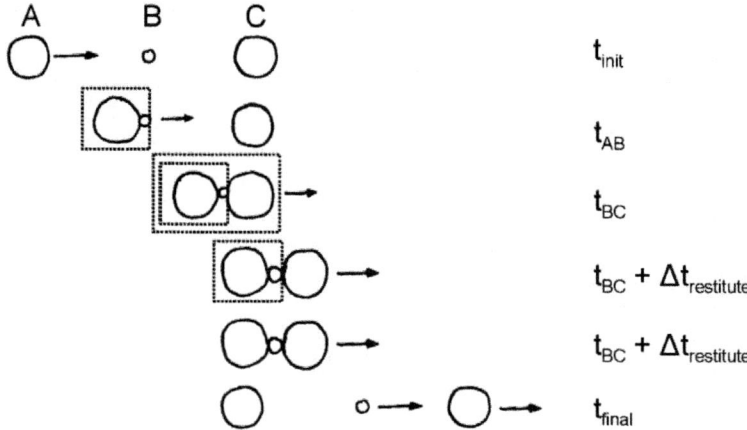

Fig. 6. A scenario demonstrating an approximation that addresses the problem of simultaneous and nearly-simultaneous collisions. Particles form loaded groups for durations of $\Delta t_{restitute}$, during which time they act as single bodies.

At time t_{AB} in Figure 6, particles A and B collide, form a loaded group, and proceed with matching velocities. Suppose that this loaded group did not encounter any other particle. In that case, at time $t_{AB} + \Delta t_{restitute}$, particles A and B would separate or "restitute". But that does not happen, as at time t_{BC} while A and B are still loaded, they encounter particle C. The impulse delivered to C depends not on the mass of B, but rather on the mass of A and B added together. It is the temporary accumulation of mass that tends to increase the momentum transferred per collision, and thus reduces the number of collisions.

It is necessary that particles in a loaded group restitute in the opposite order from that in which they loaded. After all three particles form a loaded group at t_{BC}, particles A and B may no longer separate at time $t_{AB} + \Delta t_{restitute}$. The loaded group remains intact until $t_{BC} + \Delta t_{restitute}$ instead, at which point particle B separates from C. The result of the B-C restitution is calculated with the masses of A and B still combined. After the B-C restitution is complete, but also at the simulated time $t_{BC} + \Delta t_{restitute}$, particles A and B finally separate.

Calculations pertaining to loading and restitution are relatively simple. Suppose particle A, in a loaded group with mass M_A, collides with particle B, in a loaded group with mass M_B. After loading, which takes place immediately, the new velocity of every particle in both groups is v_{load}.

$$v_{load} = \left(1 + \frac{M_B}{M_A}\right)^{-1} \cdot v_A + \left(1 + \frac{M_A}{M_B}\right)^{-1} \cdot v_B$$

At a later time, when the two loaded groups separate in the restitution phase of the collision, an impulse Δp_{AB} is applied between the groups. It value is obtained by taking the impulse calculated from either (3) or (5), and subtracting the impulse that was effectively applied when the velocities where changed to v_{load}.

$$\Delta p_{AB} = \Delta p - \left(\frac{1}{M_A} + \frac{1}{M_B}\right)^{-1} \cdot v_{AB}$$

The proposed approximation can dramatically reduce the number of collisions in a simulation, even if $\Delta t_{restitute}$ is very small. For the scenario involving three particles in a line, with the outside two particles being a thousand times more massive than the middle particle, the approximation reduced 70 elastic collisions to only four. If $\Delta t_{restitute} = 0$, loading and restitution occur back-to-back and the approximation is effectively canceled.

3.5   External Impulses

The impulses described in Sections 3.1 through 3.4 above arise from interactions between at least two particles. The novelty of the TPS method lies in the fact that these impulses alone are capable of predicting processes of deformation. However, if the particles in a TPS model are influenced by no other factor whatsoever, then one is limited to simulating objects moving deterministically in a gravity-free vacuum. In order to capture Brownian motion, drag forces, and other effects influencing the motion of biological objects, it is necessary to introduce external impulses into a TPS model. An "external impulse" is an instantaneous change in momentum that may be applied to a single particle any point in time during a simulation.

Different methods and models may be used to calculate the timing, the directions, and the magnitudes of external impulses. A realistic TPS model of a biological system may combine external impulses of many different types. We recommend that modelers at least incorporate a type of external impulse we refer to as a "random impulse": a momentum change of randomized magnitude and randomized direction applied to a particle each time a randomized time interval expires. From a practical perspective, random impulses prevent the kinetic energy in a TPS model from converging to zero due to the energy losses associated with particle collisions. From a physical perspective, random impulses may represent Brownian motion, variability in electric potential fields or fluid pressure, or interactions with otherwise unrepresented biological entities.

An external force can be represented by a sequence of external impulses. To incorporate a gravitational acceleration of g, for example, one may apply downward impulses of magnitude $m_A \cdot g \cdot \Delta t_g$ to each particle of mass m_A at regular time intervals of Δt_g. One can also use external impulses to model the drag force exerted on a particle by the surrounding fluid. Suppose that a particle of radius r_A and velocity v_A is immersed in a fluid with a dynamic viscosity of μ_{fluid} and a velocity of v_{fluid}. One could approxiate the drag force on the particle using Stoke's law, and apply external impulses of $6 \cdot \pi \cdot \mu_{fluid} \cdot r_A \cdot (v_{fluid} - v_A) \cdot \Delta t_d$ at time intervals of Δt_d. In small-scale biological models, it would in many instances be reasonable to assume $\boldsymbol{v}_{fluid}$ to be zero.

When an external impulse $\Delta \boldsymbol{p}$ is applied to particle of mass m_A and velocity $\boldsymbol{v}_A$, the new velocity $\boldsymbol{v}_A'$ is calculated as follows.

$$\boldsymbol{v}_A' = \boldsymbol{v}_A + \frac{\Delta \boldsymbol{p}}{m_A}$$

For the sake of simplicity, we recommend that each external impulse be associated with a single particle, and that $\Delta \boldsymbol{p}$ be calculated without accounting for surrounding particles that may be temporarily in the same loaded group (see Section 3.4). Once $\Delta \boldsymbol{p}$ is calculated, then if the particle happens to be in a loaded group of mass M_A, the velocity of every particle in the group is changed from $\boldsymbol{v}_{load}$ to $\boldsymbol{v}_{load}'$.

$$\boldsymbol{v}_{load}' = \boldsymbol{v}_{load} + \frac{\Delta \boldsymbol{p}}{M_A}$$

3.6   Particle Species

In a TPS model, it is useful to define several distinct species of particles. Certain properties are chosen for each species individually, and certain properties are associated with each combination of two species.

Associated with each species is the mass m of every particle of that species. We also include several parameters pertaining to the random impulses described in Section 3.5. For each species that we wish to exhibit random motion, we select an average time interval τ_{RI} between random impulses. The actual intervals are sampled from an exponential distribution during the simulation. We obtain the magnitude of each impulse from a gamma distribution, which requires a shape parameter k_{RI} and an average momentum value μ_{RI}.

There are five parameters associated with each pair of species: the blocking distance $\Delta u_{blocking}$, the tethering distance $\Delta u_{tethering}$, and the coefficients $c_{rebound}$, $c_{retract}$, and $c_{revolve}$. Suppose there are three species A and B and C, for example. There are then six combinations of two species: A-A, B-B, C-C, A-B, B-C, and C-A. Thus we would have a total of 30 parameters. In practice, the selection of most of these parameters turns out to be trivial, for we will likely want most pairs of species to remain untethered. If two A particles cannot become tethered, then the A-A tethering distance is ∞ and the coefficients $c_{retract}$ and $c_{revolve}$ are irrelevant. Also, instead of choosing blocking distances for each pair of species, we can choose radii for each individual species and add them together to obtain the blocking distances.

4   Tethered Particle System Models

Here we present TPS models of two types of deformable biological structures: vesicle clusters and membranes.

4.1 Vesicle Clusters

An action potential, a signal that propagates along the axon of a nerve cell, will ultimately arrive at a presynaptic nerve terminal. Inside this compartment are tens or hundreds of neurotransmitter-containing sacs called synaptic vesicles [17]. These vesicles bind with a certain type of protein, called synapsin, to form clusters. Vesicles can also become docked at the active zone on the membrane of the compartment. When an action potential arrives, these docked vesicles may release neurotransmitters and trigger an action potential in an adjacent neuron. Our focus in this section is on simulations that capture the dynamics of vesicle clusters as deformable structures, the formation of these clusters, and the manner in which they congregate at the active zone.

Consider a TPS model consisting of particles of three different species: V, S, and D. Each V particle represents a vesicle. Synapsins, being dimers, are represented by pairs of tethered S particles. A D particle is a docking site, a mobile location in the active zone of the membrane on which a vesicle may become docked. Such a model was used in the simulation of Figure 7, which shows two vesicles (large spheres) surrounded by synapsins (dimers) and docking sites (small spheres). The vesicles are both tethered to opposite ends of a synapsin.

Fig. 7. A snapshot of a simulation showing two vesicles (V particles), three synapsins (pairs of S particles) and seven docking sites (D particles)

The tethering of particles is governed by the following rules.

− A V particle and another V particle may never be tethered (vesicles do not bind to one another directly).
− An S particle and another S particle may be tethered at the start of a simulation; if they are not tethered at the beginning, they will never become tethered, and if they are initially tethered, they will never detach (an S particle and its tethered counterpart represent one synapsin).
− An S particle and a V particle will become tethered if they collide, if the S particle is not already tethered to a vesicle (at most two vesicles may bind to a two-particle synapsin), and if the V particle is not already tethered to the S particle's counterpart (we do not allow both ends of a synapsin to bind to the same vesicle).

- A D particle and another D particle may never be tethered.
- A D particle and a V particle will become tethered if they collide, if the D particle is not already tethered to another V particle, and if the V particle is not already tethered to another D particle (vesicles and docking sites pair up).
- A D particle and an S particle may never be tethered.

Table 1 lists the blocking and tethering distances we selected for V, S, and D particles. As indicated in the table, approaching docking site and vesicle particles collide and rebound at 25 nm. If tethered and separating, they retract at 30 nm. The distances are chosen to reflect the sizes of actual structures. The diameter of a vesicle is roughly 40 nm, for example, the value used for the vesicle-vesicle blocking distance. Note that a blocking distance of zero indicates that blocking collisions never occur between those species, whereas a tethering distance of ∞ indicates that tethering collisions never occur.

Table 1. Blocking and tethering distances for particles representing vesicles, synapsins, and docking sites

Particle Species Pair	Blocking Distance $\Delta u_{blocking}$ (nm)	Tethering Distance $\Delta u_{tethering}$ (nm)
V-V	40	∞
S-S	2.5	7.5
S-V	22.5	25
D-D	10	∞
D-V	25	30
D-S	0	∞

The masses of V and S particles are chosen to be roughly proportional to their volumes, whereas each D particle is assigned a relatively high mass for its size to account for resistance in the membrane. The rebounding, retraction, and revolution coefficients are selected such that a considerable amount of kinetic energy is lost when vesicles, synapsins, and docking sites collide. Random impulses are applied to all three of these types of particles to maintain a certain level of kinetic energy in the entire system.

In order to model the formation, disruption, and motion of vesicle clusters, it is necessary to constrain the V, S, and D particles to a region representing the presynaptic nerve terminal compartment. The simplest way to achieve this is to model the nerve cell membrane as a rigid sphere. This is done by adding two particles to the model, one with species M and one with species Z. Both of these particles are given infinite mass, which ensures that they remain stationary. The particle of species M, representing the membrane, is tethered to all V (vesicle), S (synapsin), and D (docking site) particles.

For the sake of convenience, we introduce the parameter r_M to approximate the radius of the compartment, r_V to approximate that of a vesicle, and r_S

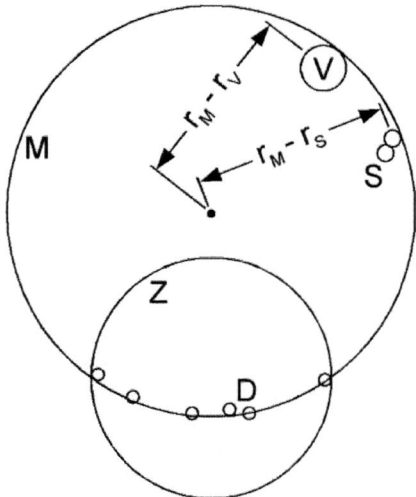

Fig. 8. A diagram illustrating the relationships between five particle species

to approximate the radius of half of a synapsin. As illustrated in Figure 8, an
M-V (membrane-vesicle) tethering distance of $r_M - r_V$ keeps vesicles in the
compartment, and an M-S (membrane-synapsin) tethering distance of $r_M - r_S$
does the same for synapsins. Because vesicles and synapsins move freely within
the compartment, the M-V and M-S blocking distances are both zero.

The D particles, representing docking sites, must be constrained to the spherical surface representing the cell membrane. The M-D blocking and tethering
distances are therefore both chosen to be near r_M, with $\Delta u_{tethering}$ slightly
greater than $\Delta u_{blocking}$. Another constraint on the docking sites is that they
must all be located in that region of the membrane known as the active zone.
Hence, all D particles are tethered to the Z particle shown in Figure 8. With the
exception of this tethering, the Z particle has no influence on any other particle.

Figure 9 shows four snapshots of a simulation of a presynaptic terminal of a
nerve cell. Vesicles and smaller synapsins move inside the semi-transparent M
particle, while docking sites move slowly along the bottom of the membrane.
The Z particle that constrains the docking sites is invisible.

Initially, the location of each vesicle and synapsin is randomized within the
spherical compartment. None of the vesicles are initially tethered to synapsin.
After the simulation begins, the tethering of colliding V and S particles leads
to the formation of vesicle clusters. These clusters, which begin to take shape
in Figures 9b, and 9c, grow fewer in number but larger in size as the simulation
progresses. The tethering of V and D particles constrains some of these clusters
to the membrane. In Figure 9d, all of the vesicles have gathered in a single
cluster at the active zone. Synaptic vesicles are typically observed in similar
membrane-bound clusters in real presynaptic nerve terminals.

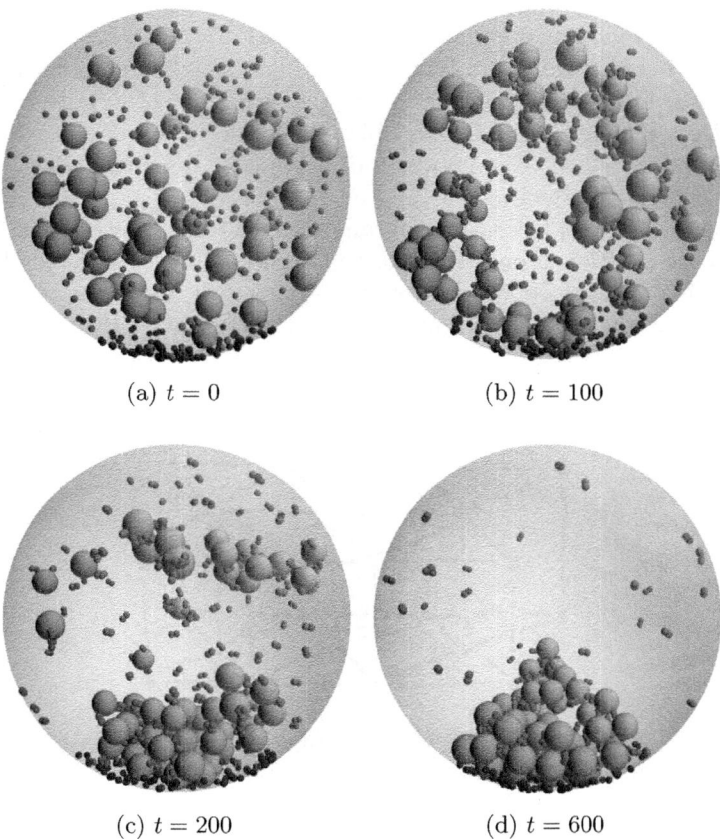

(a) $t = 0$ (b) $t = 100$

(c) $t = 200$ (d) $t = 600$

Fig. 9. Snapshots of a simulation of a presynaptic nerve terminal with a rigid spherical membrane. With a randomized initial distribution, vesicles form clusters that eventually congregate at the active zone at the bottom of the membrane.

Experimental results presented in [4] suggest that synapsin helps to maintain a number of vesicles in the vicinity of the active zone, which in turn increases the chance that a sequence of action potentials will be transmitted from one neuron to the next. Dr. James J. Cheetham, a biologist at Carleton University, uses TPS models like the one in Figure 9 to investigate this theory. By performing numerous simulations with different numbers of vesicles and synapsins, the availablility of vesicles at the active zone can be quantified as a function of synapsin concentration. The research involves an iterative process in which the TPS model is repeatedly improved, and successive sets of simulation results are compared with experimental data. One improvement made to date is the inclusion of action potentials, which cause certain tethered vesicles and synapsins to separate from one another over a period of time. New vesicle-synapsin bonds form after each action potential subsides.

4.2   Deformable Membranes

Although the rigid spherical membrane of Section 4.1 is likely adequate for a number of investigations involving vesicle clusters, the representation of deformable membranes may help capture the dynamics of a presynaptic nerve terminal on a larger scale. Deformable membranes may also prove useful for models of entire nerve cells, networks of nerve cells, tissues, blood vessels, and possibly even large organs.

A simple way to represent a membrane with a TPS is illustrated in Figure 10. Particles are positioned on a surface, and each particle is tethered the nearest neighboring particles. To avoid excessively-sharp folds and other anomalous features, a membrane should have at least two layers; that is, there should be two or more parallel surfaces of particles, and corresponding particles on adjacent surfaces should be tethered together.

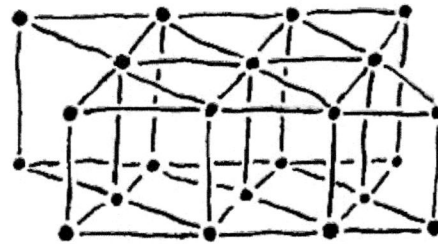

Fig. 10. An illustration of how deformable membranes may be represented. Dots are particles, and lines indicate pairs of tethered particles.

Particles on a membrane surface need not be coplanar, and need not be arranged in the triangular grid pattern shown in Figure 10. One alternative is demonstrated in Figure 11a, which shows an initially spherical membrane deforming in response to an impact with an initially downward-moving particle. The particles in the membrane were arranged in two concentric icosahedral grids, each constructed by iteratively interpolating the edges of a 20-sided regular polyhedron. In Figure 11b, this deformable icosahedral structure is used as a nerve cell membrane enclosing a presynaptic compartment. The membrane is coerced into a pear-like shape through the selection of initial particle velocities.

Several challenges can be identified by observing the results of the Figure 11b simulation. First, the edges of the underlying 20-sided polyhedron tend to protrude from the membrane as it deforms. A more random distribution of particles would reduce this effect. Another challenge pertains to resolution. The nerve cell membrane in the figure is too thick, but reducing its width would require greater numbers of particles and computations. Because intracellular fluids are essentially incompressible, maintaining the interior volume of a closed deformable membrane in a TPS model is yet another challenge. It might be possible to correct a changing interior volume at regular time intervals using external

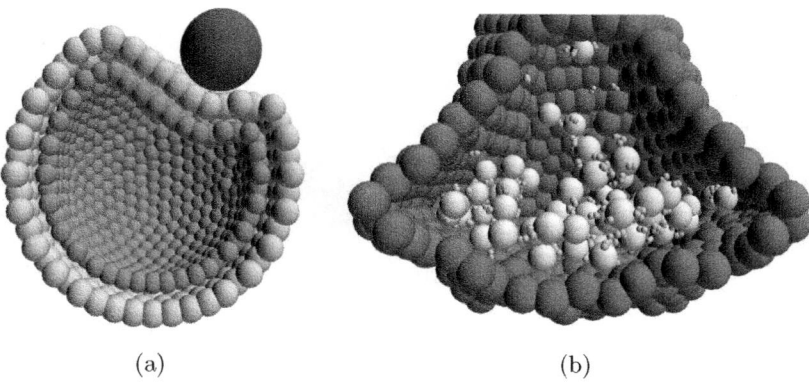

(a)   (b)

Fig. 11. On the left, an initially-spherical deformable membrane suffers an impact. On the right, a presynaptic nerve terminal is simulated with the same deformable membrane. The front half of the membrane is not shown in either snapshot, and the outer membrane layer is not shown on the right.

impulses, though the calculation of those impulses would require the use of a fluid dynamics algorithm in conjunction with the TPS.

Figure 12 shows an effort to simulate a square biological membrane or soft tissue clamped along two opposing edges. The membrane has two layers of particles, and each particle on the inside is tethered to the four adjacent particles in the same layer and one particle in the opposite layer. All particles along the two clamped edges are assigned a mass of ∞ and an initially-zero velocity, rendering them immobile.

Gravity is incorporated in the Figure 12 model via downward external impulses applied to each mobile particle at regular time intervals. As a result of these impulses, the initially flat membrane in Figure 12a is starting to sag in Figure 12b. In Figure 12c the membrane exhibits a wave-like pattern as it responds to internal tethering collisions triggered by the initial fall. Small nondeterministic ripples appear in the membrane as a consequence of two sources of randomness; the order in which particles receive gravitational impulses is randomized, as is the order in which simultaneous collisions are resolved. The gravity-induced waves have mostly subsided after 48 time units, as shown in Figure 12e. Shortly after, a falling particle impacts the membrane and produces the small ridge in Figure 12f.

The simulation of Figure 12 reveals a case for which the TPS method should not be used: a structure composed of numerous tethered particles subject to sustained opposing external forces. The opposing forces in the Figure 12 model include the downward force of gravity and the upward reaction force acting on the membrane along the stationary edges. While a macroscopic soft tissue will lengthen as the applied force increases, a TPS deformable membrane will reach its maximum length and stop stretching. If, for whatever reason, the force of gravity were to increase by an order of magnitude, the membrane in Figure 12 would sag faster but no further. The Figure 12 simulation is also computationally

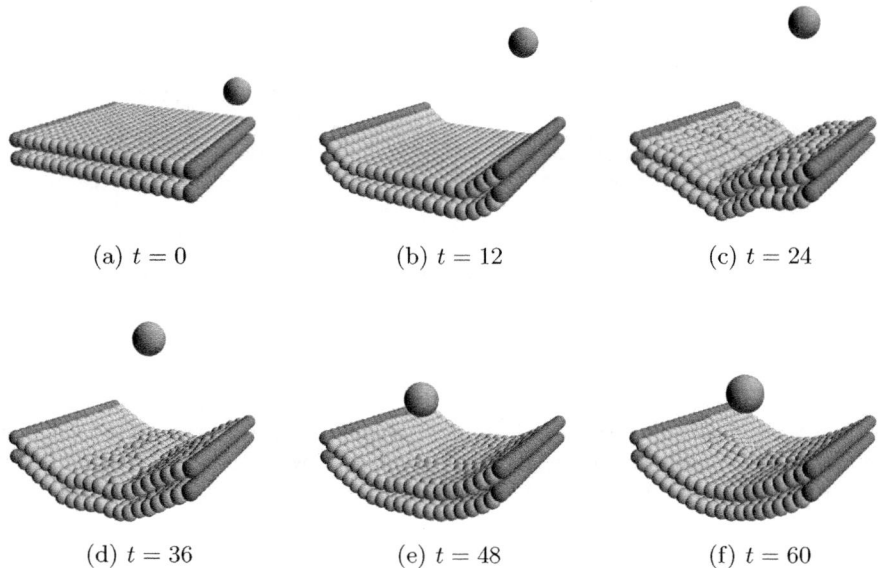

(a) $t = 0$ (b) $t = 12$ (c) $t = 24$

(d) $t = 36$ (e) $t = 48$ (f) $t = 60$

Fig. 12. A square membrane clamped along two edges responds first to gravity, then to an impact with a projectile. Note that the projectile is moving upwards in Figures 12a, 12b, and 12c, then downwards in 12d and 12e. In 12f, it is moving up after colliding with the membrane.

inefficient. After the membrane has sagged, many pairs of particles remain near or at their tethering distances. These stable contacts cause collisions to occur at an extremely high rate, slowing the simulation to a crawl.

5   Conclusion

The TPS method described and demonstrated in this paper provides convincing evidence that impulse-based methods can be used to simulate the dynamics of deformable structures. The new method is very similar to that of [14], yet contradicts the assertion that such impulse-based methods require models to be comprised of only rigid bodies. The TPS requires only analytic calculations, alleviates the need for regular time intervals, and is particularly promising for simulations of small-scale self-assembling deformable biological structures. External impulses may be added to a TPS model to produce random motion, to apply drag forces, or to account for other factors influencing the dynamics of biological objects. We have demonstrated the application of the TPS to vesicle clusters and membranes.

As mentioned in Section 4.2, the method as currently defined does not appear to be useful for models of complex macroscopic structures subject to sustained and opposing external forces. The simulation of a clamped membrane subject to gravity, shown in Section 4.2's Figure 12, is a good example of an application requiring either an alternative method, or perhaps some future enhanced version of the TPS. In the case of small-scale self-assembling biological structures subject to random motion, like the vesicle clusters of Figure 9, stable contacts are far less likely to pose a problem.

A detailed comparison of the TPS with alternative dynamic simulation methods remains important future work. Here we speculate that the FEM would be difficult to apply to small-scale self-assembling biological structures, as rapid deformation and re-structuring would require frequent re-calculation of mass and stiffness matrices. Mass-spring-damper systems suffer from the threat of instability, though it is possible to address this problem with constraints as done in the method of [2] and [6]. Recall from Section 2.2 that, although it is described as "impulse-based", the [2]/[6] method requires new trajectories to be computed for each node in a deformable structure at regular time intervals. The TPS simulates deformation with impulses applied not at regular intervals, but rather in response to collisions. The [2]/[6] method seems to be the more computationally efficient, as the number of collisions in a TPS simulation can be extremely high. The TPS is appealing in that all calculations are analytic; there is no need for the iterative algorithm of [2]/[6] that repeatedly re-calculates trajectories until all constraints are satisfied within an arbitrary tolerance level.

Impulse-based methods like the TPS are compelling in large part because they alleviate the need to choose regular time intervals. In the case of a biological system, the selection of an appropriate time interval would be complicated by the fact that interacting biological entities may differ in size and momentum by many orders of magnitude. A suitable interval for one entity may be too large or too small for another.

Acknowledgments

We thank Dr. James J. Cheetham, from the Department of Biology at Carleton University, for providing expertise in the biology of presynaptic nerve terminals and numerous suggestions on how to model them.

References

1. Baraff, D.: Analytical Methods for Dynamic Simulation of Non-penetrating Rigid Bodies. Computer Graphics 23(3), 223–232 (1989)
2. Bender, J., Bayer, D.: Impulse-based simulation of inextensible cloth. In: Proceedings of The International Conference on Computer Graphics and Visualization (IADIS), Amsterdam, Netherlands (2008)
3. Brown, J., Sorkin, S., Bruyns, C., Latombe, J.-C., Stephanides, M., Montgomery, K.: Real-time simulation of deformable objects: Tools and application. Computer Animation, 228–236 (2001)
4. Camilli, P.D.: Keeping synapses up to speed. Nature 375, 450–451 (1995)
5. Conti, F., Khatib, O., Baur, C.: Interactive rendering of deformable objects based on a filling sphere modeling approach. In: Proceedings of the IEEE International Conference on Robotics and Automation (ICRA), Taipei, Taiwan (2003)

6. Diziol, R., Bender, J., Bayer, D.: Volume Conserving Simulation of Deformable Bodies. In: Proceedings of Eurographics, Munich, Germany (2009)

7. Gibson, S.F.F., Mirtich, B.: A Survey of Deformable Modeling in Computer Graphics. Mitsubishi Electric Reasearch Laboratories (1997), www.merl.com

8. Guess, T.M., Maletsky, L.P.: Computational modelling of a total knee prosthetic loaded in a dynamic knee simulator. Medical Engineering & Physics 27(5), 357–367 (2005)

9. Jansson, J., Vergeest, J.S.M.: A discrete mechanics model for deformable bodies. Computer-Aided Design 34(12), 913–928 (2002)

10. Keiser, R., Müller, M., Heidelberger, B., Teschner, M., Gross, M.: Contact Handling for Deformable Point-Based Objects. In: Proceedings of the Vision, Modeling, and Visualization Conference (VMV), Stanford, CA, USA (2004)

11. Kitano, H.: Computational systems biology. Nature 420, 206–210 (2002)

12. Klein, M.L., Shinoda, W.: Large-Scale Molecular Dynamics Simulations of Self-Assembling Systems. Science 321(5890), 798–800 (2008)

13. Mirtich, B., Canny, J.: Impulse-based Simulation of Rigid Bodies. In: Proceedings of the 1995 Symposium on Interactive 3D Graphics (SI3D), Monterey, California, United States (1995)

14. Mirtich, B.V.: Impulse-based Dynamic Simulation of Rigid Body Systems. PhD. University of California at Berkeley, Berkeley (1996)

15. Moore, M., Wilhelms, J.: Collision Detection and Response for Computer Animation. Computer Graphics 22(4), 289–298 (1988)

16. Moore, P., Molloy, D.: A Survey of Computer-Based Deformable Models. In: Proceedings of the International Machine Vision and Image Processing Conference (IMVIP), Maynooth, Ireland (2007)

17. Südhof, T.C., Starke, K.: Pharmacology of Neurotransmitter Release. Springer, Heidelberg (2008)

18. Tagawa, K., Hirota, K., Hirose, M.: Impulse Response Deformation Model: an Approach to Haptic Interaction with Dynamically Deformable Object. In: Proceedings of the Symposium on Haptic Interfaces for Virtual Environment and Teleoperator Systems (HAPTICS), Alexandria, VA, USA (2006)

19. Wang, T., Pan, T.-W., Xing, Z.W., Glowinski, R.: Numerical simulation of rheology of red blood cell rouleaux in microchannels. Physical Review E (Statistical, Nonlinear, and Soft Matter Physics) 79(4), 041916+ (2009)

Delay Stochastic Simulation of Biological Systems: A Purely Delayed Approach

Roberto Barbuti, Giulio Caravagna, Andrea Maggiolo-Schettini,
and Paolo Milazzo

Dipartimento di Informatica, Università di Pisa
Largo Pontecorvo 3, 56127 Pisa, Italy
{barbuti,caravagn,maggiolo,milazzo}@di.unipi.it

Abstract. Delays in biological systems may be used to model events for which the underlying dynamics cannot be precisely observed. Mathematical modeling of biological systems with delays is usually based on Delay Differential Equations (DDEs), a kind of differential equations in which the derivative of the unknown function at a certain time is given in terms of the values of the function at previous times. In the literature, delay stochastic simulation algorithms have been proposed. These algorithms follow a "delay as duration" approach, which is not suitable for biological systems in which species involved in a delayed interaction can be involved at the same time in other interactions. We show on a DDE model of tumor growth that the delay as duration approach for stochastic simulation is not precise, and we propose a simulation algorithm based on a "purely delayed" interpretation of delays which provides better results on the considered model. Moreover, we give a formal definition of a stochastic simulation algorithm which combines both the delay as duration approach and the purely delayed approach.

1 Introduction

Biological systems can often be modeled at different abstraction levels. A simple event in a model that describes the system at a certain level of detail may correspond to a rather complex network of events in a lower level description. The choice of the abstraction level of a model usually depends on the knowledge of the system and on the efficiency of the analysis tools to be applied to the model.

Delays may appear in models of biological systems at any abstraction level, and are associated with events whose underlying dynamics either cannot be precisely observed or is too complex to be handled efficiently by analysis tools. Roughly, a delay may represent the time necessary for the underlying network of events to produce some result observable in the higher level model.

Mathematical modeling of biological systems with delays is mainly based on delay differential equations (DDEs), a kind of differential equations in which the derivative of the unknown function at a certain time is given in terms of the values of the function at previous times. In particular, this framework is very

C. Priami et al. (Eds.): Trans. on Comput. Syst. Biol. XIII, LNBI 6575, pp. 61–84, 2011.

general and allows both simple (constant) and complex (variable or distributed) forms of delays to be modeled.

As examples of DDE models of biological systems we mention [3,15,10,14,7]. In [3,15] an epidemiological model is defined that computes the theoretical number of people infected with a contagious illness in a closed population over time; in the model a delay is used to model the length of the infectious period. In [10] a simple predator-prey model with harvesting and time delays is presented; in the model a constant delay is used based on the assumption that the change rate of predators depends on the number of prey and predators at some previous time. Finally, models of tumor growth [14] and of HIV cellular infection [7] have been presented and analyzed by using DDEs.

Models based on DDEs, similarly to their simplest versions based on ordinary differential equations (ODEs), may be studied either analytically (by finding the solution of the equations, equilibria and bifurcation points) or via approximated numerical solutions. However, for complex real models analytical solutions are often difficult or impossible to be computed, whereas their approximated numerical solution is more feasible.

Models based on differential equations, although very useful when dealing with biological systems involving a huge number of components, are not suitable to model systems in which the quantity of some species is small. This is caused by the fact that differential equations represent discrete quantities with continuous variables, and when quantities are close to zero this becomes a too imprecise approximation. In these cases a more precise description of systems behavior can be obtained with stochastic models, where quantities are discrete and stochastic occurrence of events is taken into account.

The most common analysis technique for stochastic models is stochastic simulation that, in the case of models of biological systems without delays, often exploits Gillespie's Stochastic Simulation Algorithm (SSA) of chemical reactions [9], or one of its approximated variants [8,6]. In recent years, the interest for stochastic delayed processes increased [13]. In [2] a Delay Stochastic Simulation Algorithm (DSSA) has been proposed, this algorithm gives an interpretation as durations to delays. The delay associated with a chemical reaction whose reactants are consumed (i.e. are not also products) is interpreted as the duration of the reaction itself. Such an interpretation implies that the products of a chemical reaction with a delay are added to the state of the simulation not at the same time of reactants removal, but after a quantity of time corresponding to the delay. Hence, reactants cannot be involved in other reactions during the time modeled by the delay.

We argue that the interpretation of delay as duration is not always suitable for biological systems. We propose a simple variant of the DSSA in which reactants removal and products insertion are performed together after the delay. This corresponds to a different interpretation of delays, that is the delay is seen as the time needed for preparing an event which happens at the end of the delay. An example of a biological behavior which can be suitably modeled by this interpretation is mitosis. Cell mitosis is characterized by a pre–mitotic phase and by a mitotic phase (cell division). The pre–mitotic phase prepares the division of the cell, when a cell undergoes the mitotic process, the pre–mitotic phase can be seen as a delay before the real cell division. During the pre–mitotic phase the cell can continue to interact with the environment, for example it can die. The DSSA in [2] cannot model this interactions because the reactants (in this case the cell itself) are removed at the beginning of reaction and the products are added at its end (that is after the delay).

In this paper we start by recalling the definition of DDEs and a DDE model of tumor growth [14]. Then, we give a stochastic model of the considered tumor growth example and simulate it by using the DSSA introduced in [2] and based on an interpretation of delays as durations. Subsequently, we propose a new "purely delayed" interpretation of delays and, consequently, a new variant of the DSSA that we apply to the considered tumor growth example. Although this new DSSA permits to have more precise results than the DSSA in [2] and it has a very easy implementation, there exist some scenarios in which this version of the algorithm does not work properly. Hence, we define a more precise version of it which requires a much more complex implementation. Before drawing our conclusions, we give a formal definition of a stochastic simulation algorithm which combines both the delay as duration approach and the purely delayed approach in its most precise definition.

2   Delay Differential Equations (DDEs)

The mathematical modeling of biological systems is often based on Ordinary Differential Equations (ODEs) describing the dynamics of the considered systems in terms of variation of the quantities of the involved species over time.

Whenever phenomena presenting a delayed effect are described by differential equations, we move from ODEs to *Delay Differential Equations* (DDEs). In DDEs the derivatives at current time depend on some past states of the system. The general form of a DDE for $X(t) \in \mathbb{R}^n$ is

$$\frac{dX}{dt} = f_x(t, X(t), \{X(t') : t' \leq t\}),$$

The simplest form of DDE considers *constant delays*, namely consists of equations of the form

$$\frac{dX}{dt} = f_x(t, X(t), X(t - \sigma_1), \dots, X(t - \sigma_n))$$

with $\sigma_1 > \dots > \sigma_n \geq 0$ and $\sigma_i \in \mathbb{R}$. This form of DDE allows models to describe events having a fixed duration. They have been used to describe biological systems in which events have a non-negligible duration [3,15,10] or in which a sequence of simple events is abstracted as a single complex event associated with a duration [14,7].

In what follows we recall an example of DDE model of a biological system that we shall use to compare delay stochastic simulation approaches.

2.1   A DDE Model of Tumor Growth

Villasana and Radunskaya proposed in [14] a DDE model of tumor growth that includes the immune system response and a phase-specific drug able to alter the natural course of action of the cell cycle of the tumor cells.

The cell cycle is a series of sequential events leading to cell replication via cell division. It consists of four phases: G_1, S, G_2 and M. The first three phases (G_1, S, G_2) are called interphase. In these phases, the main event which happens is the replication of DNA. In the last phase (M), called mitosis, the cell segregates the duplicated sets of chromosomes between daughter cells and then divides. The duration of the cell cycle depends on the type of cell (e.g a human normal cell takes approximately 24 hours to perform a cycle).

The model in [14] considers three populations of cells: the immune system, the population of tumor cells during cell cycle interphase, and the population of tumor cells during mitosis. A delay is used to model the duration of the interphase, hence the model includes a delayed event that is the passage of a tumor cell from the population of those in the interphase to the population of those in the mitotic phase. In the model the effect of a phase-specific drug, able to arrest tumor cells during the mitosis, is studied. Such a drug has a negative influence also on the survival of cells of the immune system.

In this paper we study a simplified version of the model (presented in subsection 4.1.2 of [14]), where the effects of the immune response and of the drug are not taken into account. The simplified model, which considers only tumor cells (both in pre-mitotic and mitotic phases), consists of the following DDEs:

$$\frac{dT_I}{dt} = 2a_4T_M - d_2T_I - a_1T_I(t - \sigma) \qquad T_I(t) = \phi_0(t) \text{ for } t \in [-\sigma, 0]$$

$$\frac{dT_M}{dt} = a_1T_I(t - \sigma) - d_3T_M - a_4T_M \qquad T_M(t) = \phi_1(t) \text{ for } t \in [-\sigma, 0]$$

Function $T_I(t)$ denotes the population of tumor cells during interphase at time t, and function $T_M(t)$ denotes the tumor population during mitosis at time t. The terms d_2T_I and d_3T_M represent cell deaths, or apoptosis. The constants a_1 and a_4 represent the phase change rates from interphase to mitosis (a_1) and back (a_4). In the following we shall denote with d the rate at which mitotic cells disappear, namely $d = d_3 + a_4$.

We assume that cells reside in the interphase at least σ units of time; then the number of cells that enter mitosis at time t depends on the number of cells that entered the interphase at least σ units of time before. This is modeled by the terms $T_I(t - \sigma)$ in the DDEs. Note that each cell leaving the mitotic phase produces two new cells in the T_I population (term $2a_4T_M$). In the model the growth of the tumor cell population is obtained only through mitosis, and is given by the constants a_1, a_4, and σ which regulate the pace of cell division. The delay σ requires the values of T_I and T_M to be given also in the interval $[-\sigma, 0]$: such values are assumed to be constant in the considered interval, and hence equal to the values of T_I and T_M at time 0.

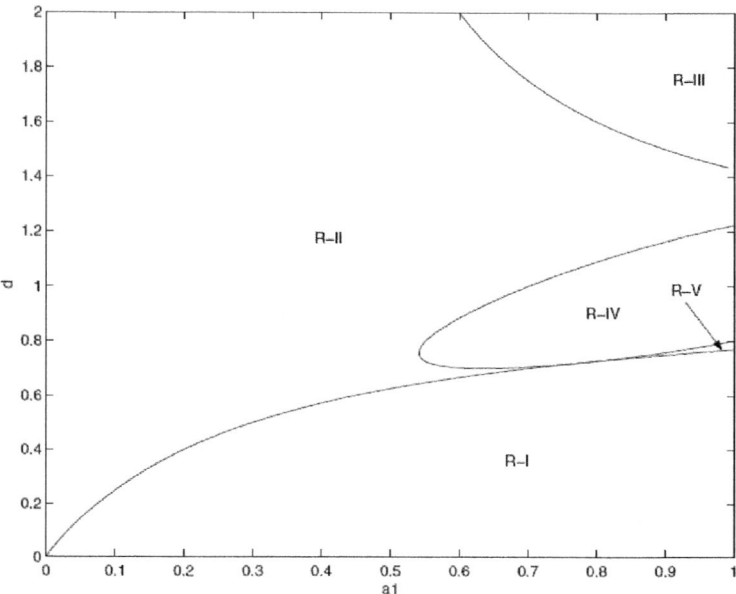

Fig. 1. The regions which describe the different behaviors of the DDE model by varying parameters a_1 and d (picture taken from [14])

The analytic study of the DDEs constituting the model gives $(0,0)$ as unique equilibrium. In Figure 1 (taken from [14]) some results are shown of the study of the model by varying a_1, d and σ and by setting the parameters a_4 and d_2 to 0.5 and 0.3, respectively. Figure 1 shows five regions.

When $\sigma = 0$, the region in which the tumor grows is R-I, while in the other regions the tumor decays.

When the delay is present $(\sigma > 0)$, the growth region is essentially unaltered, but the decay is split into regions in which the tumor has different behaviors: in regions R-II $\cup$ R-IV the tumor still decays, but in regions R-III $\cup$ R-V, when the value of σ is sufficiently large, the equilibrium becomes unstable. This is shown in Figures 2 and 3.

Figure 2 describes the behavior of the model, obtained by numerical solutions, inside the regions R-I, R-II, R-III, and R-IV, when $\sigma = 1$. Actually, we considered the point $(0.6, 0.6)$ in R-I, the point $(0.4, 1.0)$ in R-II, the point $(1.0, 1.8)$ in R-III, the point $(0.8, 0.8)$ in R-IV and an initial state consisting in 10^5 tumor cells in the interphase and 10^5 tumor cells in mitosis. We shall use always these parameters in the rest of the paper. In the figure, we can observe that, while the tumor grows in region R-I, it decays in all the other regions.

Figure 3 describes the behavior of the model when $\sigma = 10$. In regions R-I and R-IV the tumor has the same behavior as before. In region R-II it decays after some oscillations, while in region R-III it expresses an instability around the equilibrium. However, remark that values of T_M and T_I under 0 are not realistic, and, as we will see in the following, they cannot be obtained by stochastic simulations.

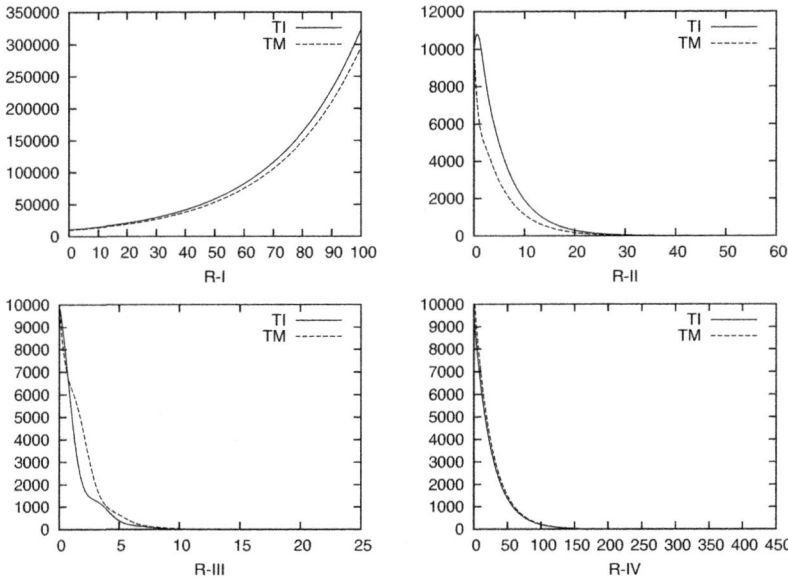

Fig. 2. Results of the numerical solution of the DDE model with $\sigma = 1$ for the regions described in Figure 1. On the x-axis time is given in *days* and on the y-axis is given the *number of cells*.

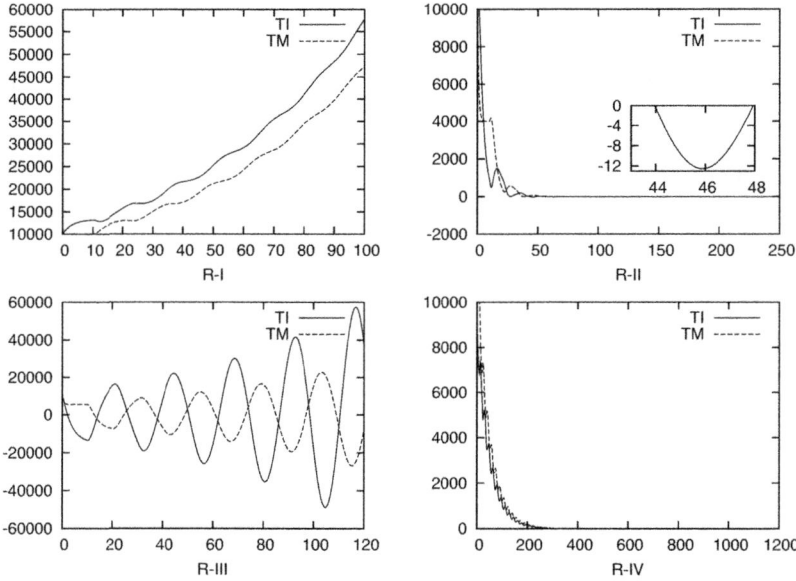

Fig. 3. Results of the approximated numerical simulation of the DDE model with $\sigma = 10$ for the regions described in Figure 1. On the x-axis time is given in *days* and on the y-axis is given the *number of cells*.

3 Delay Stochastic Simulation

In this section we present algorithms for the stochastic simulation of biological systems with delays. Firstly, we introduce a well–known formulation of one of these algorithms, and we analyze the results of the simulations of the stochastic model equivalent to the one presented in the previous section. Secondly, we propose a variant of this algorithm and we compare the results of the simulations done by using this algorithm with those of the simulation done by using the original one.

All the simulations and the algorithms that we are going to present in this section have been implemented in the software tool DelaySim. This tool, available at http://www.di.unipi.it/msvbio, has been written in Java.

3.1 The Delay as Duration Approach (DDA)

In [2] Barrio *et al.* introduced a *Delay Stochastic Simulation Algorithm* (DSSA) by adding delays to Gillespie's Stochastic Simulation Algorithm (SSA) [9]. The algorithm has been used to explain more carefully than with DDE models the observed sustained oscillations in the expression levels of some proteins.

In order to recall the definition of the algorithm in [2] we consider a well–stirred system of *molecules* of N chemical *species* $\{S_1, \ldots, S_N\}$ interacting through M chemical *reaction channels* $\mathcal{R} = R_1, \ldots, R_M$. We assume the volume and the temperature of the system to be constant. We denote the number of molecules of species S_i in the system at time t with $X_i(t)$, and we want to study the evolution of the *state vector* $\mathbf{X}(t) = (X_1(t), \ldots, X_N(t))$, by assuming that the system was initially in some state $\mathbf{X}(t_0) = \mathbf{x}_0$.

A reaction channel R_j is characterized mathematically by three quantities. The first is its *state–change vector* $\nu_j = (\nu_{1j}, \ldots, \nu_{Nj})$, where ν_{ij} is defined to be the change in the S_i molecular population caused by one R_j reaction; let us denote each state–change vector ν_j as a the composition of the state–change vector for reactants, ν_j^r, and the state–change vector for products, ν_j^p, noting that $\nu_j = \nu_j^r + \nu_j^p$. For instance, given two species A and B, a reaction of the form $A \to B$ is described by the vector of reactants $(-1, 0)$, by the vector of products $(0, 1)$ and by the state–change vector $(-1, 1)$; differently, a reaction of the form $A \to A + B$ is described by the vector of reactants $(-1, 0)$, by the vector of products $(1, 1)$, and by the state–change vector $(0, 1)$.

The second characterizing quantity for a reaction channel R_j is its *propensity function* $a_j(\mathbf{x})$; this is defined, accordingly to [9], so that, given $\mathbf{X}(t) = \mathbf{x}$, $a_j(\mathbf{x})dt$ is the probability of reaction R_j to occur in state $\mathbf{x}$ in the time interval $[t, t + dt]$. As stated in [9], the propensity function can be defined as follows

$$a_j(\mathbf{x}) = k \cdot \prod_{i=1}^{N} \binom{X_i(t)}{|\nu_{i,j}^r|}$$

where $k \in \mathbb{R}$ denotes the kinetic function of reaction R_j and $|\nu_{i,j}^r|$ denotes the absolute value of the i-th coordinate of vector ν_j^r. This probabilistic definition finds its justification in physical theory.

Algorithm **DSSA with "delays as duration approach"**

1. Initialize the time $t = t_0$ and the system state $\mathbf{x} = \mathbf{x}_0$.
2. Evaluate all the $a_j(\mathbf{x})$ and their sum $a_0(\mathbf{x}) = \sum_{j=1}^{M} a_j(\mathbf{x})$;
3. Given two random numbers r_1, r_2 uniformly distributed in the interval $[0; 1]$, generate values for τ and j in accordance to

$$\tau = \frac{1}{a_0(\mathbf{x})} \ln(\frac{1}{r_1}) \qquad \sum_{i=1}^{j-1} a_i(\mathbf{x}) < r_2 \cdot a_0(\mathbf{x}) \leq \sum_{i=1}^{j} a_i(\mathbf{x})$$

   (A) If delayed reaction R_k [a] is scheduled at time $t + \tau_k$ and $\tau_k < \tau$
      (A1) If $R_k \in \mathcal{R}_{nc}$ then update $\mathbf{x} = \mathbf{x} + \nu_k$ and $t = t + \tau_k$;
      (A2) If $R_k \in \mathcal{R}_c$ then update $\mathbf{x} = \mathbf{x} + \nu_k^p$ and $t = t + \tau_k$;
   (B) else:
      (B1) If $R_j \in \mathcal{R}_{nd}$ then update $\mathbf{x} = \mathbf{x} + \nu_j$ and $t = t + \tau$;
      (B2) If $R_j \in \mathcal{R}_{nc}$, schedule R_j at time $t + \sigma_j + \tau$ and set time to $t + \tau$;
      (B3) If $R_j \in \mathcal{R}_c$, schedule R_j at time $t + \sigma_j + \tau$, update $\mathbf{x} = \mathbf{x} + \nu_k^r$ and set time to $t + \tau$;
4. go to step 2.

---

[a] This is the reaction with minimum τ_k, hence the first to complete.

Fig. 4. The DSSA with "delays as duration approach" proposed in [2]

The other characterizing quantity is a constant delay defined by a real number $\sigma \geq 0$. Following Barrio *et al.*, we classify reactions with delays into two categories: non-consuming reactions, where the reactants are also products (e.g. $A \to A + B$), and consuming reactions, where some of the reactants are consumed (e.g. $A \to B$). Throughout the paper, we denote the set of non-consuming reactions with delay by $\mathcal{R}_{nc}$, the set of consuming reactions with delay by $\mathcal{R}_c$, and the reactions without delays by $\mathcal{R}_{nd}$; notice that $\mathcal{R} = \mathcal{R}_{nc} \cup \mathcal{R}_c \cup \mathcal{R}_{nd}$ and $\mathcal{R}_{nc}$, $\mathcal{R}_c$ and $\mathcal{R}_{nd}$ are pair–wise disjoint.

By adding delays to the SSA, Barrio *et al.* provide a method to model the firing of a reaction with delay based on the previously given classification. Formally, given a system in state $\mathbf{X}(t) = \mathbf{x}$, let us denote with τ the stochastic time quantity computed as in the SSA representing the putative time for next reaction to fire. Let us assume to choose to fire a non-consuming reaction with delay (a reaction from set $\mathcal{R}_{nc}$); then the reaction is scheduled at time $t + \sigma + \tau$, where σ is the delay of the reaction. Furthermore, the clock is increased to the value $t + \tau$ and the state does not change. On the contrary, if a consuming reaction with delay (a reaction from set $\mathcal{R}_c$) is chosen to fire, then its reactants are immediately removed from the state $\mathbf{x}$, the insertion of the products is scheduled at time $t + \sigma + \tau$, and, finally, the clock is increased to the value $t + \tau$. Reactions from set $\mathcal{R}_{nd}$ (non–delayed reactions) are dealt with exactly as in the SSA. The DSSA by Barrio *et al.* is given in Figure 4.

We discuss now on the scheduling of the reactions with delay. When a non-consuming reaction is chosen, the algorithm does not change state, but simply schedules the firing of the reaction at time $t + \sigma_j + \tau$ (step $(B2)$). The reaction will complete its firing (reactants and products will be removed and inserted, respectively) when performing steps (A) and $(A1)$.

Differently, as regards consuming reactions, the removal of the reactants is done at time instant t (step $(B3)$) preceding the time instant of insertion of the products (steps (A) and $(A2)$), namely the time at which the insertion is scheduled, $t + \sigma_j + \tau$. Notice that the removed reactants cannot have other interactions during the time interval $[t, t + \sigma_j + \tau)$.

As the reactants cannot have other interactions in the time quantity passing between the removal of the reactants and the insertion of the products, then this quantity can be seen as a duration needed for the reactants to exclusively complete the reaction. Since the approach of Barrio *at al.* gives this interpretation of delays we shall call it "delays as duration approach" (DDA).

As regards the handling of the scheduled events (step (A) of the algorithm), if in the time interval $[t; t + \tau)$ there are scheduled reactions, then τ is rejected and the scheduled reaction is handled. At this step the algorithm implicitly assumes R_k to be the scheduled action with the minimum τ_k. Among all the others which could be chosen the choice of the one with the minimum τ_k is quite intuitive since this will be the first to complete. Since generating random numbers is a costly operation, other authors defined variants of the DSSA that avoid rejecting τ in the handling of scheduled reactions [5,1]. However, the interpretation of the delays used to define these variants is the same as that of Barrio *et al.*.

This interpretation of delays may not be precise for all biological systems. In particular, it may be not precise if in the biological system the reactants can have other interactions during the time window modeled by the delay. The tumor growth system we have recalled in Section 2.1 is an example of these systems. In fact, while tumor cells are involved in the phase change from interphase to mitosis (the delayed event) they can also die.

We applied the DSSA by Barrio *at al.* (we refer to the simulations done by applying this DSSA as DDA simulations) to a chemical reaction model corresponding to the DDE model of tumor growth recalled in Section 2.1. The reactions of the model are the following:

$$T_I \xrightarrow{a_1} T_M \text{ with delay } \sigma \qquad T_M \xrightarrow{a_4} 2T_I \qquad T_I \xrightarrow{d_2} \qquad T_M \xrightarrow{d_3} .$$

We have run 100 simulations for each considered parameter setting. The results of simulations with the same parameters as those considered in Figures 2 and 3 are shown in Figures 5 and 6, respectively. Actually, in the figures we show the result of one randomly chosen simulation run for each parameter setting.

Qualitatively, results obtained with DDA simulations are the same as those obtained with numerical simulation of the DDEs: we have exponential tumor growth in region R-I, tumor decay in the other regions and oscillations arise when the delay is increased. However, from the quantitative point of view we have that in the DDA simulations the growth in region R-I and the decay in the

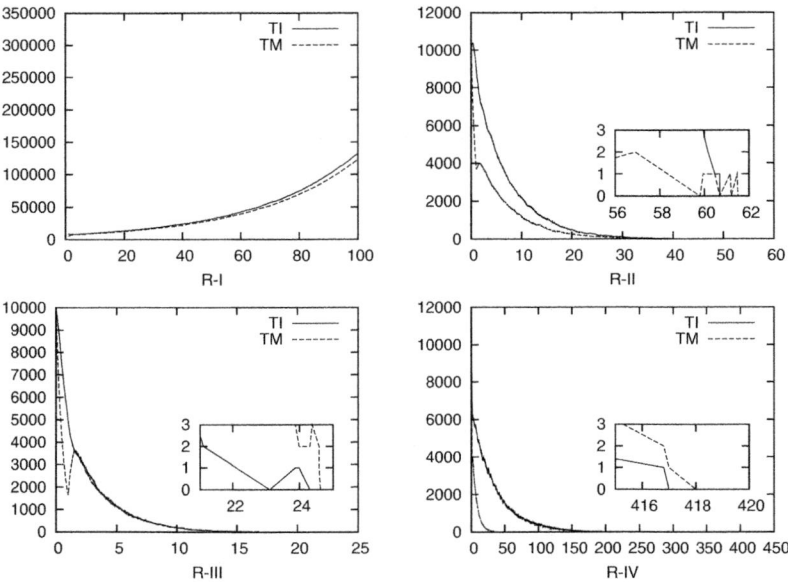

Fig. 5. DDA simulation of the stochastic model with $\sigma = 1$ for the regions described in Figure 1. On the x-axis time is given in *days* and on the y-axis is given the *number of cells*.

other regions are always slower than in the corresponding numerical simulation of the DDEs. In fact, with $\sigma = 1$ by the numerical simulation of the DDEs we have that in region R-I after 100 days both the quantities of tumor cells in interphase and in mitotic phase are around 300000, while in the result of DDA simulations they are around 130000. In the same conditions, but with $\sigma = 10$, in the numerical simulation of the DDEs we have about 47000 tumor cells in mitosis and 57000 tumor cells in interphase, while in the DDA simulations we have about 5000 and 5500 cells, respectively. As regards the other regions, in Table 1 the average tumor eradication times obtained with DDA simulations are compared with those obtained with numerical simulation of the DDEs (in this case with "eradication" we mean that the number of tumor cells of both kinds is under the value 1). Again, we have that in DDA simulations the dynamics is slower than in the numerical simulation of the DDEs. For instance, with $\sigma = 10$, in region R-IV the time needed for eradication in the DDEs is about 41% of the time needed in the DDA (440 against 1072), in region R-II the percentage is smaller, 26% (59 against 224), and, in region R-III, it reaches 9% (12 against 126). For the same regions with $\sigma = 1$ these differences are smaller but not negligible.

3.2   A Purely Delayed Approach (PDA)

In this section we propose a variant of the DSSA based on a different interpretation of delays, namely a Stochastic Simulation Algorithm which follows a

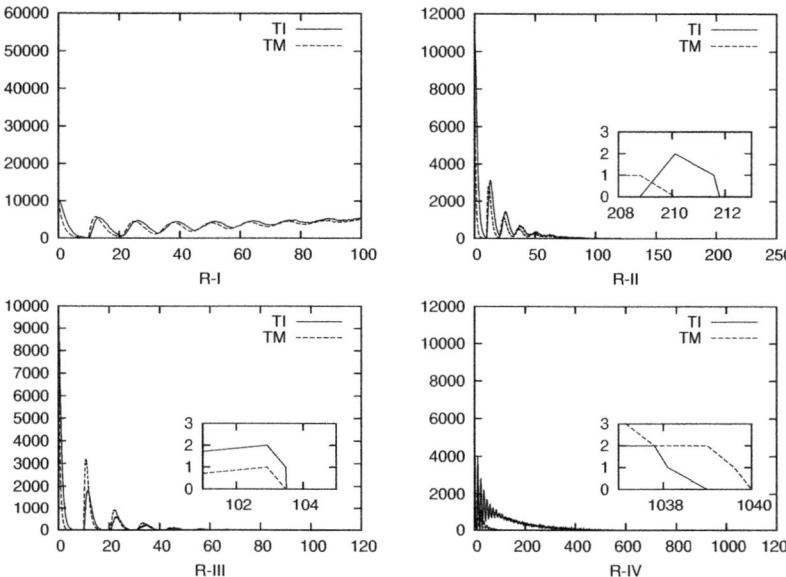

Fig. 6. DDA simulation of the stochastic model with $\sigma = 10$ for the regions described in Figure 1. On the x-axis time is given in *days* and on the y-axis is given the *number of cells*.

"purely delayed approach" (PDA). With this interpretation we try to overcome the fact that in the DDA the reactants cannot have other interactions. Furthermore, differently from Barrio *et al.*, we use the same interpretation of delays to define the method for firing both non-consuming and consuming reactions. This interpretation of delays was firstly implicitly adopted by Bratsun *et al.* in [4], to model a very simple example of protein degradation.

The approach we propose consists in firing a reaction completely when its associated scheduled events is handled, namely removing its reactants and inserting its products after the delay. The fact that we simply schedule delayed reactions without immediately removing their reactants motivates the terminology of "purely delayed". Notice that non-consuming reactions are handled in the same way by DDA and PDA.

In this interpretation of delays it may happen that, when handling a scheduled reaction (again assuming to pick the one with the minimum τ_k as in the DDA since this is the first to complete), the reactants may not be present in the current state. In fact, they could have been destroyed or transformed by other interactions happened after the scheduling. In this case, the scheduled reaction has to be ignored. To formalize this, we know that a reaction R_j can be applied only if its reactants are all present in the current state of the simulation. Algebraically this corresponds to the fact that $\nu_j^r \prec \mathbf{x}$ where ν_j^r is the state–change vector of the reactants of reaction R_j, the system is described by $\mathbf{x}$ and $\prec$ is the ordering relation defined as $\forall i = 1, \ldots, N. \ -\nu_{ij}^r \leq X_i(t)$. In order to verify that a scheduled reaction can effectively fire, it will be sufficient to check whether

Algorithm **DSSA with "purely delayed approach"**

1. Initialize the time $t = t_0$ and the system state $\mathbf{x} = \mathbf{x}_0$.
2. Evaluate all the $a_j(\mathbf{x})$ and their sum $a_0(\mathbf{x}) = \sum_{j=1}^{M} a_j(\mathbf{x})$;
3. Given two random numbers r_1, r_2 uniformly distributed in the interval $[0, 1]$, generate values for τ and j in accordance to

$$\tau = \frac{1}{a_0(t)} \ln(\frac{1}{r_1}) \qquad \sum_{i=1}^{j-1} a_i(\mathbf{X}(t)) < r_2 \cdot a_0(t) \le \sum_{i=1}^{j} a_i(\mathbf{X}(t))$$

   (a) If delayed reaction R_k [a] is scheduled at time $t + \tau_k$ and $\tau_k < \tau$ and $\nu_k^r \prec \mathbf{x}$, then update $\mathbf{x} = \mathbf{x} + \nu_k$ and $t = t + \tau_k$;
   (b) else, schedule R_j at time $t + \sigma_j + \tau$, set time to $t + \tau$;
4. go to step 2.

---

[a] This is the reaction with minimum τ_k, hence the first to complete.

Fig. 7. The DSSA with "purely delayed approach"

Table 1. Average eradication times given in *days* for DDE model, DDA and PDA stochastic models. For the stochastic models the entries represent the sample of 100 simulations.

	DDEs	DDA Simulation	PDA Simulation
R-II with $\sigma = 1.0$	50	64	51
R-II with $\sigma = 10.0$	59	224	67
R-III with $\sigma = 1.0$	15	29	17
R-III with $\sigma = 10.0$	12	126	20
R-IV with $\sigma = 1.0$	238	302	214
R-IV with $\sigma = 10.0$	440	1072	248

this condition holds. The formal definition of the DSSA with PDA is given in Figure 7.

As for the DDA, we have run 100 simulations of the stochastic model of tumor growth for each considered parameter setting. The results of simulations (we refer to these simulations as PDA simulations) with the same parameters as those considered in Figures 2 and 3 are shown in Figures 8 and 9, respectively. Actually, in the figures we show the result of one randomly chosen simulation run for each parameter setting.

Qualitatively, results obtained with PDA simulations are the same as those obtained with numerical simulation of the DDEs (and with DDA simulations). From the quantitative point of view we have that in the PDA simulations the growth in region R-I with $\sigma = 1$ is almost equal to the corresponding numerical simulation of the DDEs (about 300000 tumor cells in both mitosis and interphase after 100 days, we recall that the DDA had reached values around 130000). On the contrary, with $\sigma = 10$, the difference between DDEs and PDA is higher: we have about 22000 tumor cells in interphase against 57000 for the DDEs and 5500 for the DDA, and 16000 tumor cells in mitosis against 47000 for the DDEs and 5000 for the DDA.

As regards the other regions, in Table 1 the average tumor eradication times obtained with PDA simulations are compared with those obtained with numerical simulation of the DDEs (again, in this case with "eradication" we mean that the number of tumor cells of both kinds is under the value 1). In PDA simulations the dynamics is generally slower than in the numerical simulation of the DDEs but it is faster than the DDA one. With $\sigma = 10$, in region R-IV the time needed for eradication in the PDA is smaller than the one in the DDEs (248 days against 440, DDA is 1072). In region R-II the values are: 67 days for the PDA and 59 days for the DDEs, DDA is 224. In region R-III values are: 20 days for the PDA, 12 days for the DDEs, and 126 days for DDA.

It is important to remark that differences between delay stochastic simulation results and numerical solutions of DDEs are also influenced by the initial conditions. The numerical solution of the DDEs assumes the initial population to be constant and greater than zero in the time interval $[-\sigma, 0]$. This allows delayed event to be enabled in the time interval $[0, \sigma]$. Both variants of the DSSA start to schedule delayed events from time 0, hence delayed reactions can fire only after the time σ. This results, when σ is great enough, in a behavior that is, in general, delayed with respect to that given by the DDEs.

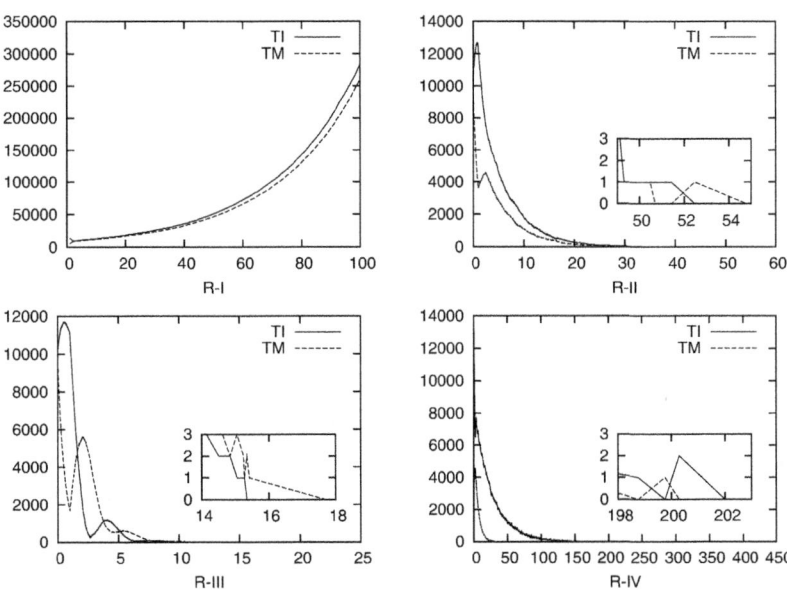

Fig. 8. PDA simulation of the stochastic model with $\sigma = 1$ for the regions described in Figure 1. On the x-axis time is given in *days* and on the y-axis is given the *number of cells*.

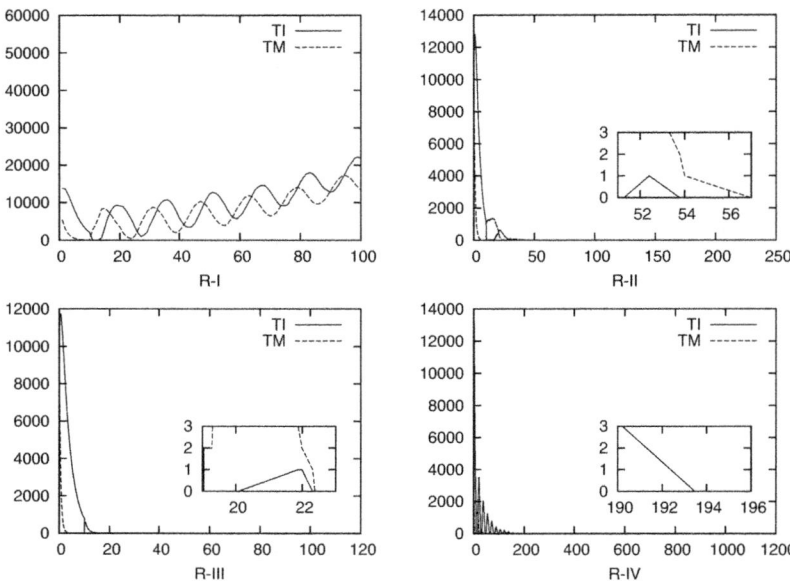

Fig. 9. PDA simulation of the stochastic model with $\sigma = 10$ for the regions described in Figure 1. On the x-axis time is given in *days* and on the y-axis is given the *number of cells*.

Now, even if for this particular model this PDA definition is enough to justify the introduction of simulation techniques different from the DDA one, we must make some considerations about the PDA algorithm we proposed. In particular, there are some scenarios in which the algorithm does not work properly. We will go through these scenarios via some examples. For instance, consider a system described by the following initial state and chemical reaction:

$$\mathbf{X}(t_0) = (1,\ 0) \qquad\qquad A \xrightarrow{k,\sigma} B$$

where the initial state contains one single molecule A and the only reaction is the one transforming a molecule A in a molecule B with a kinetic constant k and a delay $\sigma > 0$. When applying the PDA algorithm to simulate this system it is easy to observe that the algorithm may over-schedule the firing of the reaction. This could happen because the reaction has a delay and, dependently on the value σ and on the random numbers generated by the algorithm, between time t_0 and the first firing of the scheduled reactions, the PDA may schedule an arbitrary amount of times the reaction. However, after applying the first scheduled reaction at time $t + \varphi$, the system state becomes the vector $\mathbf{X}(t_0 + \varphi) = (0,\ 1)$ and, consequently, all the other scheduled reactions are not applicable anymore. This situation is not incorrect in fact, as expected, just one molecule A is transformed in a molecule B, but it is computationally unpleasant since it schedules a lot of reactions that will be never performed.

Also, it is possible to define, on top of this scenario, a new model such that the behavior of the PDA becomes incorrect. Imagine, for instance, to have the same scenario enriched with a reaction which produces, by an external unbounded source, molecules of type A. As there is no way of tracking the time since a molecule is in the system, then there is no way of preventing to apply a scheduled rule to a molecule A which is not in the current state of the system by, at least, σ time units. In this enriched scenario it may be the case that molecules A just appearing in the state by the firing of the new reaction, may be used to perform over-scheduled reactions and this is, obviously, incorrect.

Consequently, even if the PDA is, for some biological systems, a better candidate than the DDA algorithm by Barrio *et al.*, it needs to be properly tuned to avoid to simulate incorrect behaviors of the modeled system. The next section will be devoted to the definition of a variant of the PDA in order to face these issues. The more precise variant of the PDA will also be extended to obtain an algorithm that integrates the two approaches.

3.3   The PDA with Markings

In order to get a correct version of the PDA we consider a solution based on a marking of molecules. This variant of the PDA, in the following named "marked Purely Delay Approach" (mPDA), is based on the idea of assigning, to each molecule of the system, a marking which permits the identification of the molecules involved in any scheduled reaction. On one side, this will fix the liabilities of the PDA approach but, as it is intuitive, it will be computationally much more expensive than the PDA.

In order to define the mPDA, we still assume the framework we used to introduce the PDA where the state vector $\mathbf{X}(t)$ describes the time evolution of a set of molecules belonging to n chemical species and $\mathcal{R}$ denotes a set of of chemical reaction channels.

In what follows we describe the mPDA, whose definition will be given in Figure 10.

Marking the molecules. The marking of molecules is based on the use of natural numbers as identifiers. In order to get a clear marking policy we classify the molecules of the system. Firstly, the molecules are classified in species, hence a system is described by a set of species $\mathcal{S} = \{\Sigma_1, \ldots, \Sigma_n\}$ which defines the type of molecules we are considering. Furthermore, any Σ_i denotes a set of molecules such that

$$\Sigma_i = \{S_{N_1}^{i,1}, \ldots, S_{N_{n_i}}^{i,n_i}\}$$

where $S_N^{i,j}$ is a single molecule belonging to species $\Sigma_i \in \mathcal{S}$, with a unique identifier $j \in \mathbb{N}$ and concurrently performing the reactions in $N \subset \wp(\{z | R_z \in \mathcal{R}\})$, a set of identifiers of the reactions which are present in the current model. Notice that, for any molecule of the model, we carry much more information than the one we had in the PDA. In particular, for any molecule we can exactly know which reactions it is concurrently performing and, in this context, this

Algorithm **DSSA with "marked purely delayed approach"**

1. Initialize the time $t = t_0$, build the initial marking w.r.t definition (2) by using the input initial state $\mathbf{x}_0$.
2. Evaluate all the $a_j(\mathbf{x})$ w.r.t. definition (3), define $a_0(\mathbf{x}) = \sum_{j=1}^{M} a_j(\mathbf{x})$;
3. Given two random numbers r_1, r_2 uniformly distributed in the interval $[0, 1]$, generate values for τ and j in accordance to

$$\tau = \frac{1}{a_0(t)} \ln(\frac{1}{r_1}) \qquad \sum_{i=1}^{j-1} a_i(\mathbf{x}) < r_2 \cdot a_0(t) \leq \sum_{i=1}^{j} a_i(\mathbf{x})$$

   (a) If delayed reaction R_k [a] is scheduled at time $t + \tau_k$ and $\tau_k < \tau$, then:
       - update the event list w.r.t. definition (7);
       - update the marking w.r.t definitions (6), (8), (9) and (10);
       - set time to $t = t + \tau_k$.
   (b) else:
       - choose, w.r.t. definition (4), the set of reactants E that will be modified by reaction R_j;
       - update the marking w.r.t. definition (5);
       - schedule the triple $(j, t + \sigma_j + \tau, E)$;
       - set time to $t + \tau$.
4. go to step 2.

---

[a] This is the reaction with minimum τ_k, hence the first to complete.

Fig. 10. The DSSA with "marked purely delayed approach"

means that there is an instance of reaction consuming that molecule which is currently scheduled in the event list. As these sets change during the simulation of a system, we may denote by $\mathcal{S}(t)$ and $\Sigma_i(t)$ the set of species and the set of molecules of species Σ_i at time t, respectively.

The marking of molecules requires to discuss the use of vector $\mathbf{X}(t)$ which, as in the PDA, will be used to observe the state changes due to the reactions firing. The construction of the state vector $\mathbf{X}(t)$ is slightly changed, with respect to the PDA, by the introduction of this marking notation. In particular, we define $\mathbf{X}(t)$ as

$$\mathbf{X}(t) = (|\Sigma_1(t)|, \ldots, |\Sigma_n(t)|) \tag{1}$$

where $|\Sigma_i(t)|$ denotes the cardinality of the set $\Sigma_i(t)$. Notice that $|\Sigma_i(t)|$ represents the number of molecules of species Σ_i at time t, exactly as the element $X_i(t)$ of $\mathbf{X}(t)$ in the definition of the PDA.

As regards $\mathbf{X}(t_0)$, given an initial input state $\mathbf{x}_0$ for the mPDA, a proper initial marking for the system to simulate has to be computed. Let us assume $\mathbf{x}_0 = (X_1(t_0), \ldots, X_n(t_0))$, the set of species can be defined as $\mathcal{S}(t_0) = \{\Sigma_1(t_0), \ldots, \Sigma_n(t_0)\}$ where

$$\forall i = 1, \ldots, n. \ \Sigma_i(t_0) = \{S_\emptyset^{i,1}, \ldots, S_\emptyset^{i, X_i(t_0)}\}. \tag{2}$$

This construction creates n sets of species types $\Sigma_i(t_0)$ and, for each of them, it creates $X_i(t_0)$ molecules, each one with a different identifier, which are performing no reactions (at time t_0 the event list is empty, hence their set subscript is the empty set). This guarantees that the molecules are correctly marked and, hence, distinguishable.

In general, at any step of computation, the mPDA algorithm may modify some of the sets $\Sigma_i(t) \in \mathcal{S}(t)$. The discussion about how the mPDA modifies the sets in $\mathcal{S}(t)$, accordingly to the time–evolution of the simulated system, is presented in the forthcoming subsections.

Evaluating the propensity functions. The firing of a reaction is the result of a probabilistic choice based on the propensity function of the reaction, evaluated in the current state of the simulation. In order to explain how the mPDA works, we recall the assumption which is at the basis of the definition of this PDA algorithm. The molecules can perform multiple reactions in parallel, but each molecule can be involved in each reaction at most once at a time. The first reaction to finish interrupts the others running in parallel and involving the same molecule. In order to avoid over-scheduling phenomena it is to be ensured that propensity function of a reaction depends only on the occurrences of reactants that are not yet involved in the same reaction.

Let us assume that we have to evaluate the propensity function of a reaction $R_z : M \xrightarrow{k,\sigma} P \in \mathcal{R}$ such that R_z transforms a multiset of molecules M in a multiset of molecules P with a kinetic constant $k \in \mathbb{R}$. More precisely, let us assume M to be a multiset of the form $\{(1, n_1), \ldots, (w, n_w)\}$, namely reaction R_z transforms, for any $j = 1, \ldots, w$, a number of n_j molecules of species Σ_j. Notice that this corresponds to a set representation of the state–change vector for reactants.

As we want to take into consideration only the molecules in the current state of the simulation which are not already involved in any scheduled firing of reaction R_z, then we have to filter those that are candidate for being used, if any. Let us denote by $[\Sigma_i(t), z]$ the set of identifiers of molecules belonging to species $\Sigma_i(t)$ which have to be considered in the evaluation of the propensity function of R_z, namely the set

$$[\Sigma_i(t), z] = \{j \mid S_N^{i,j} \in \Sigma_i(t) \wedge z \notin N\}.$$

Notice that this set is obtained by considering all the molecules in the current system, and by filtering them on the basis of the marking information that the mPDA stores in $\mathcal{S}(t)$. In the PDA this set could not have been defined.

Given $\mathbf{X}(t) = \mathbf{x}$, the propensity function $a_z(\mathbf{x})$ must consider only those molecules required by M which are not already performing reaction R_z, hence it can be defined as follows:

$$a_z(\mathbf{x}) = k \cdot \prod_{(i,n_i) \in M} \binom{|[\Sigma_i(t), z]|}{n_i} \tag{3}$$

where $|[\Sigma_i(t), z]|$ denotes the cardinality of the set $[\Sigma_i(t), z]$. This definition of the function $a_z(\mathbf{x})$ is such that mPDA propensity functions compute, in general, strictly smaller values than the PDA ones. Again, the PDA cannot distinguish the molecules which are performing a reaction from those which are not.

Scheduling a reaction to fire. Whenever the propensity functions have been evaluated, for any $R_z \in \mathcal{R}$, accordingly to the definition (3), the index of the reaction to fire can be chosen with the same policy used in the PDA. However, having a marking of molecules, the mPDA has a further level of choice to determine to which molecules the reaction will be applied.

To clarify this, as an example consider a system with two distinct molecules of the same type and both available for being consumed by a reaction. Whenever the mPDA decides to fire that reaction, it has to choose to which of the two molecules the reaction will be applied. This further choice is required by the mPDA because it stores individual information about the molecules and, hence, there exist two different destination markings that the system may reach. Notice that, as the PDA abstracted these informations, it did not perform this further choice.

In order to define this further probabilistic choice, assume the mPDA has chosen to schedule the firing of the reaction $R_z : \{(1, n_1), \ldots, (w, n_w)\} \xrightarrow{k,\sigma} P$ introduced in the previous section. The mPDA stores in the event list the same information of the PDA, namely the index of the reaction, z, and the time in which it will fire, some $t + \tau + \sigma$ if t is the current time, τ is the putative time for next reaction as computed in the PDA and σ is the delay of the reaction. Together with this information, the mPDA stores in each element of the event list a set of labels E representing the identifiers of the molecules which will be consumed by the reaction, when handled. The set E contains pairs of natural numbers and is such that if $(i, j) \in E$ then the molecule $S_N^{i,j} \in \Sigma_i(t)$ is involved in the reaction. The set E is built by considering the molecules which can effectively perform reaction R_z. Formally, for all $(i, n_i) \in M$, we choose n_i molecules from the set $[\Sigma_i, z]$. Each molecule is chosen with probability $|[\Sigma_i(t), z]|^{-1}$, hence the probability of choosing a set E, with a system at time t, denoted by $P(E, t)$, is defined as

$$P(E,t) = \prod_{(i,n_i)\in M} \binom{|[\Sigma_i(t), z]|}{n_i}^{-1} \tag{4}$$

The mPDA updates the system clock to a value $t + \tau$, stores the triple $(z, t + \tau + \sigma, E)$ in the event list and changes the marking of the molecules belonging to the set E. The marking is updated to store the information that the molecules in E are performing reaction R_z. This will guarantee that, when evaluating the propensity function for reaction R_z in the next time, the molecules in E will not be counted again, as expected. The updated set $\mathcal{S}(t + \tau)$, built by modification of the set $\mathcal{S}(t)$ satisfies the following proposition

$$\forall i = 1, \ldots, n. \Sigma_i(t + \tau) = \{S_N^{i,j} \in \Sigma_i(t) \mid (i, j) \notin E\} \cup$$

$$\{S_{N \cup \{z\}}^{i,j} \in \Sigma_i(t) \mid S_N^{i,j} \in \Sigma_i(t) \wedge (i, j) \in E\}. \tag{5}$$

Intuitively, any molecule in $\Sigma_i(t)$ that has not been assigned to the firing of reaction R_z is simply copied in $\Sigma_i(t + \tau)$. Differently, all the molecules assigned to this firing of R_z, are copied in $\Sigma_i(t + \tau)$ with the index z added to their set of concurrently running reactions.

Handling a scheduled reaction. When the mPDA decides, with the system at time t, to handle a scheduled reaction R_z it finds, as information in the event list, a triple (z, t', E) where z is the identifier of the reaction to fire, t' is the time to which the clock must be set and E is the set of identifiers of the molecules which will be consumed by the reaction. It is guaranteed, by construction, that the molecules denoted by the set E are still present in the current state of the simulation. Hence, differently from the PDA, the condition $\nu_z^r \prec \mathbf{x}$ has not to be checked at this time.

The scheduled reaction is applied, as expected, by using the same policy of the PDA, namely the reactants are removed and the products are inserted. However, the mPDA must perform some additional operations to keep the marking of the molecules correct.

First of all, let us assume the set $E = \{(s_1, l_1), \ldots, (s_m, l_m)\}$, then all the molecules denoted by these labels in $\mathcal{S}(t)$ must not be present anymore in the set $\mathcal{S}(t')$, built by the mPDA to represent the markings after the application of reaction R_k. In particular, for any $j = 1, \ldots, m$, the molecule $S_N^{s_j, l_j}$ must be removed from the proper set in $\mathcal{S}(t')$. To define this, we start by defining the following sets

$$\forall i = 1, \ldots, n.\ \Sigma_i(t') = \Sigma_i(t) \setminus \{S_N^{i,j} \in \Sigma_i(t) \mid (i,j) \in E\} \tag{6}$$

Notice that this corresponds to remove exactly the number of reactants required by the application of the reaction R_z. Consequently, given the state vector $\mathbf{X}(t) = \mathbf{x}$ defined accordingly to (1), this new marking corresponds to a new state $\mathbf{x} - \nu_z^r$.

As regards the interruption of the concurrently running reactions which were assuming to use the reactants just consumed by reaction R_z, the mPDA performs two operations. Firstly, the mPDA interrupts these reactions by removing them from the event list and, secondly, it unlocks all the involved partners molecules, so that they may start again, in the future, the interrupted reactions.

The interruption of the scheduled reactions is trivial. Let us denote with $\mathcal{E}(t)$ the event list of the system at time t, all the reactions to be interrupted are those which contain, at least, one reactant which is consumed by reaction R_z. We denote by $\mathcal{B}(t)$ the set of reactions to be interrupted at time t, namely the set

$$\mathcal{B}(t) = \{(\tilde{w}, \tilde{t}, \tilde{E}) \in \mathcal{E}(t) \mid \tilde{E} \cap E \neq \emptyset\}.$$

Consequently, the mPDA modifies the event list $\mathcal{E}(t)$ creating a new event list $\mathcal{E}(t')$ such that

$$\mathcal{E}(t') = \mathcal{E}(t) \setminus \mathcal{B}(t). \tag{7}$$

Unlocking the partners of the interrupted reactions is less easy. First of all, when considering a generic molecule $S_N^{i,j} \in \Sigma_i(t')$, where $\Sigma_i(t')$ is a set of molecules satisfying (6), it may be the case that it is coupled to some of the events which have been interrupted in $\mathcal{E}(t')$, and these events belong to the set $\mathcal{B}(t)$. Also, it may be the case that the molecule is performing other reactions which have not been interrupted. In general, even if $w \in \{w | (w, t, E) \in \mathcal{B}(t)\}$, this does not imply that all the scheduled events referring to reaction R_w have to be interrupted. Clearly, this depends on the one-to-many correspondence between a reaction and all the related scheduled events. Hence, in order to filter the reactions which have been really interrupted for a molecule $S_N^{i,j} \in \Sigma_i(t')$, we define the set

$$\mathcal{D}(t, i, j) = \{w \mid (w, \tilde{t}, \tilde{E} \cup \{(i, j)\}) \in \mathcal{B}(t)\}.$$

The construction of the set $\mathcal{D}(t, i, j)$ is straightforward. All the reactions which have to be interrupted, with respect to molecule $S_N^{i,j}$, are only those relative to events effectively interrupted and such that the molecule was assumed to be consumed by that instance of reaction. This constraint filters any possible collision between the indexes of the reactions relative to the interrupted events and those which are performed with partners whose are not affected by the application of the scheduled reaction R_z.

After this considerations, we can formally define how the interruption of some events affects the marking of the molecules by defining these new sets

$$\forall i = 1, \ldots, n. \; \Sigma_i'(t') = \{S_{N'}^{i,j} \mid S_N^{i,j} \in \Sigma_i(t') \wedge N' = N \backslash \mathcal{D}(t, i, j)\}. \quad (8)$$

Notice that, as this definition does not modify the number of molecules present in the markings, then this marking, with respect to definition (1), still represents the vector state $\mathbf{x} - \nu_z^r$.

Finally, we discuss how the insertion of the products affects the marking of the molecules in the system, with respect to the sets just created. Let us assume that the scheduled reaction R_z creates a multiset of products $P = \{(1, n_1), \ldots, (p, n_p)\}$, namely R_z produces, for any $j = 1, \ldots, p$, a number n_j of new molecules of species Σ_j.

The creation of new objects to add to the sets $\Sigma_i'(t')$ requires to assign them new fresh identifiers respecting the uniqueness of the markings. As the marking is based on the use of natural numbers, the mPDA has an infinite set of numbers from which to choose the new identifiers. Let us denote, for a species Σ_i, the maximum among all the used identifiers appearing in $\Sigma_i'(t')$ as follows

$$\mu_i = \max\{j \mid S_N^{i,j} \in \Sigma_i'(t')\}.$$

Hence, for the set $\Sigma_i'(t')$, the creation of n_i non colliding identifiers can be obtained by choosing the n_i successors of the number μ_i. By these consideration we can define the following sets

$$\forall i = 1, \ldots, n. \; \Sigma_i''(t') = \Sigma_i'(t') \cup \{S_\emptyset^{i,\mu_i+1}, \ldots, S_\emptyset^{i,\mu_i+n_i} \mid (i, n_i) \in P\}. \quad (9)$$

Finally, the complete marking computed by the mPDA after the application of a scheduling rule is defined as

$$\mathcal{S}(t') = \{\Sigma_1''(t'), \ldots, \Sigma_n''(t')\}. \quad (10)$$

This new marking is obtained by modifying the one representing, accordingly to definition (1), the state vector $\mathbf{x} - \nu_z^R$. As this marking is built by inserting, for each species, exactly the number of product molecules of reaction R_z, then this new marking corresponds to the state vector $\mathbf{x} - \nu_z^r + \nu_z^p = \mathbf{x} + \nu_z$ which is, as expected, the resulting state of the correct application of reaction R_z.

3.4 A DSSA Combining the mPDA and the DDA

In this section we define a stochastic simulation algorithm which combines the delay as duration approach and the purely delayed approach in its most precise definition. This will allow biological phenomena that cannot be suitably dealt with by only one of the two approaches, to be studied.

The framework in which we define this DSSA, in the following denoted as Full DSSA, is a simple modification of the one in which we defined the mPDA. This requires to redefine the DSSA with DDA in a framework were markings are present. As regards the notation, we introduce two disjoint sets of possible reactions $\mathcal{R} = \mathcal{R}_D \cup \mathcal{R}_P$ where $\mathcal{R}_D$ and $\mathcal{R}_P$ are the sets of reactions that are treated with a DDA approach and a mPDA approach, respectively.

In what follows we describe the Full DSSA, whose definition will be given in Figure 11.

Marking the molecules. Marking the molecules is necessary to use the mPDA inside the Full DSSA. Clearly, the marking defined by the mPDA in Section 3.3, together with definitions (1) and (2), is still valid in the Full DSSA.

Evaluating the propensity functions. We introduced two disjoints sets of reactions, $\mathcal{R}_D$ and $\mathcal{R}_P$, in order to separate reactions whose delays have to be considered as durations from those whose delays are pure. However, it is easy to notice that, for any reaction in $R_z \in \mathcal{R}$, its propensity function can be correctly defined as in (3). This can be done because the different interpretations of the delays do not require different definitions of the propensity functions, but simply different semantics of the firings of reactions.

Despite this similarity, it is worth making a simple consideration about reactions in $\mathcal{R}_D$. Those reactions are such that, whenever started, they remove the reactants from the state of the simulation and, when the firing terminates, they add to the state their products. Hence, if a reaction of set $\mathcal{R}_D$ is performed by a molecule $S_N^{i,j} \in \Sigma_i(t)$, then all the reactions concurrently running in N have to be interrupted and, the involved partners, have to be unlocked. By this consideration it is easy to notice that $\forall i \in N.R_i \in \mathcal{R}_P$ and, hence, it holds that $\forall t > t_0.\ \forall z \in \mathcal{R}_D.\ [\Sigma_i(t), z] = \Sigma_i(t)$. Summarizing, evaluating the propensity function of a reaction from set $\mathcal{R}_D$ does not require to define the set $[\Sigma_i(t), z]$, an operation whose cost is at most linear in the size of $\Sigma_i(t)$, and, consequently, it is computationally less expensive than the evaluation of a propensity function of a reaction in $\mathcal{R}_P$.

Scheduling a reaction to fire. Reactions in $\mathcal{R}_P$ are scheduled accordingly to the definitions (4) and (5) whereas, the reactions in the set $\mathcal{R}_D$ are scheduled with a different policy.

Algorithm **Full DSSA**

1. Initialize the time $t = t_0$, build the initial marking w.r.t definition (2) by using the input initial state $\mathbf{x}_0$.
2. Evaluate all the $a_j(\mathbf{x})$ w.r.t. definition (3), define $a_0(\mathbf{x}) = \sum_{j=1}^{M} a_j(\mathbf{x})$;
3. Given two random numbers r_1, r_2 uniformly distributed in the interval $[0, 1]$, generate values for τ and j in accordance to

$$\tau = \frac{1}{a_0(t)} \ln(\frac{1}{r_1}) \qquad \sum_{i=1}^{j-1} a_i(\mathbf{x}) < r_2 \cdot a_0(t) \leq \sum_{i=1}^{j} a_i(\mathbf{x})$$

   (a) If delayed reaction R_k [a] is scheduled at time $t + \tau_k$ and $\tau_k < \tau$, then:
   - if $R_k \in \mathcal{R}_D$ update the marking w.r.t definitions (9) and (10);
   - if $R_k \in \mathcal{R}_P$ update the event list w.r.t. definition (7), update the marking w.r.t definitions (6), (8), (9) and (10);
   - set time to $t = t + \tau_k$.
   (b) else:
   - choose, w.r.t. definition (4), the set of reactants E that will be modified by reaction R_j;
   - if $R_j \in \mathcal{R}_D$ update the event list w.r.t. definition (7), update the marking w.r.t. definitions (6) and (8), schedule the pair $(j, t + \sigma_j + \tau)$;
   - if $R_j \in \mathcal{R}_P$ update the marking w.r.t. definition (5) and schedule the triple $(j, t + \sigma_j + \tau, E)$;
   - set time to $t + \tau$.
4. go to step 2.

---

[a] This is the reaction with minimum τ_k, hence the first to complete.

Fig. 11. The Full DSSA with both "delay as duration approach" and "marked purely delayed approach"

Assume that the Full DSSA wants to schedule a reaction $R_w \in \mathcal{R}_D$ at time $t + \tau + \sigma_w$. Firstly, the Full DSSA must choose the reactants to which the reaction is applied. As this choice is independent with respect to the interpretation of the delays, the set E to which the reaction will be applied can be chosen accordingly to definition (4), as in the mPDA.

Now, as the state must be modified by the removal of the reactants, the Full DSSA changes the marking accordingly to definition (6) which corresponds exactly to this operation. Furthermore, as the Full DSSA has, for all molecules in the set E, to interrupt all the reactions that they are concurrently performing, it modifies the event list accordingly to definition (7). Finally, the Full DSSA further modifies the marking accordingly to definition (8) in order to unlock the partners involved in the interrupted reactions.

The scheduling of the reaction is then performed by adding, to the event list $\mathcal{E}(t)$, a pair $(w, t + \tau + \sigma_w)$. Consequently, the event list in the case of the Full

DSSA contains some triples referring to scheduled reactions belonging to set $\mathcal{R}_\mathcal{P}$, and some pairs referring to scheduled reactions belonging to set $\mathcal{R}_\mathcal{D}$.

We remark that, in the mPDA, definitions (6), (7) and (8) were introduced when handling a scheduled reaction. The fact that for a reaction in $\mathcal{R}_\mathcal{D}$ the Full DSSA uses these definitions at the time of scheduling the reaction is due to the different interpretations of delays.

Handling a scheduled reaction. Scheduled reactions belonging to set $\mathcal{R}_\mathcal{P}$ are handled, accordingly to the mPDA, as explained in Section 3.3.

Differently, handling a reaction with a DDA approach is trivial because the major computational effort has been done when it was scheduled. Assume that the Full DSSA wants to handle a scheduled reaction described by the pair (w, t') where $R_w \in \mathcal{R}_\mathcal{D}$. In order to insert the product molecules of R_w by modifying the current marking the Full DSSA modifies $\mathcal{S}(t)$ by applying definitions (9) and (10).

4   Discussion

In the previous sections we showed two different approaches to the firing of delayed reactions. The two approaches can be conveniently used for dealing with two different classes of delayed reactions. The delay as duration approach suitably deals with reactions in which reactants cannot participate, whenever scheduled, in other reactions. On the other hand, the purely delayed approach can be conveniently used in cases in which reactants can be involved in other reactions during the delay time.

In the example we have shown, cells in the interphase, which wait for entering the mitotic phase, can be involved in another reaction, namely their death. Thus in this example the purely delayed approach seems to be more appropriate for capturing the behavior of this real system.

However, the algorithm we have defined, implemented and applied to the considered cell–growth model is rather naive, and may be incorrect in several scenarios. Consequently, we have defined a more precise algorithm, the mPDA, based on the purely delayed approach, which exploits a technique of marking of the molecules. The marking technique makes the mPDA computationally costly. As future work, we plan to study simplified versions of the mPDA to be proved correct by means of abstract interpretation techniques.

Furthermore, as there are biological systems in which, due to the heterogeneity of reactions, both the approaches should be used, we combined, in a new framework, both the duration as delay approach and the purely delayed approach with markings.

In the future, we plan to define formal languages for the definition of models with delays. These languages should be such that the time evolution of the described models is in accordance with the algorithms we proposed. As far as we know, similar notions of delay have been presented in the framework of Petri nets with time information. In particular, in Timed nets [12] a notion of delay similar to a duration appears; differently, in Time nets [11] the notion of delay corresponds to our purely delayed approach.

84   R. Barbuti et al.

References

1. Anderson, D.F.: A Modified Next Reaction Method for Simulating Chemical Systems with Time Dependent Propensities and Delays. J. Ch. Phys. 127(21), 214107 (2007)
2. Barrio, M., Burrage, K., Leier, A., Tian, T.: Oscillatory Regulation of Hes1: Discrete Stochastic Delay Modelling and Simulation. PLoS Computational Biology 2(9) (2006)
3. Beretta, E., Hara, T., Ma, W., Takeuchi, Y.: Permanence of an SIR Epidemic Model with Distributed Time Delays. Tohoku Mathematical Journal 54(2), 581–591 (2002)
4. Bratsun, D., Volfson, D., Tsimring, L.S., Hasty, J.: Delay-induced Stochastic Oscillations in Gene Regulation. PNAS 102(41), 14593–14598 (2005)
5. Cai, X.: Exact Stochastic Simulation of Coupled Chemical Reactions with Delays. J. Ch. Phys. 126, 124108 (2007)
6. Cao, Y., Gillespie, D., Petzold, L.: The Slow-scale Stochastic Simulation Algorithm. J. Ch. Phys. 122, 14116 (2005)
7. Culshaw, R.V., Ruan, S.: A Delay–differential Equation Model of HIV Infection of CD4+ T–cells. Mathematical Biosciences 165, 27–39 (2000)
8. Gillespie, D.: Approximate Accelerated Stochastic Simulation of Chemically Reacting Systems. J. Phys. Ch. 115, 1716 (2001)
9. Gillespie, D.: Exact Stochastic Simulation of Coupled Chemical Reactions. J. Phys. Ch. 81, 2340
10. Martin, A., Ruan, S.: Predator-prey Models with Delay and Prey Harvesting. J. Math. Biol. 43(3), 247–267 (2001)
11. Popova–Zeugmann, L., Heiner, M., Koch, I.: Time Petri Nets for Modelling and Analysis of Biochemical Networks. Fundamenta Informaticae 67, 149–162 (2005)
12. Ramchandani, C.: Analysis of Asynchronous Concurrent Systems by Timed Petri Nets. Massachussets Inst. Technol. Res. Rep., MAC-TR 120 (1974)
13. Schlicht, R., Winkler, S.: A Delay Stochastic Process with Applications in Molecular Biology. J. Math. Biol. 57, 613–648 (2008)
14. Villasana, M., Radunskaya, A.: A Delay Differential Equation Model for Tumor Growth. J. Math. Biol. 47, 270–294 (2003)
15. Zhanga, F., Lia, Z., Zhangc, F.: Global Stability of an SIR Epidemic Model with Constant Infectious Period. Applied Mathematics and Computation 199(1), 285–291 (2008)

Modelling Ammonium Transporters in Arbuscular Mycorrhiza Symbiosis[*]

Mario Coppo[1], Ferruccio Damiani[1], Maurizio Drocco[1], Elena Grassi[1,2], Mike Guether[3], and Angelo Troina[1]

[1] Dipartimento di Informatica, Università di Torino
{coppo,damiani,troina}@di.unito.it, maurizio.drocco@gmail.com
[2] Molecular Biotechnology Center, Dipartimento di Genetica, Biologia e Biochimica, Università di Torino
grassi.e@gmail.com
[3] Dipartimento di Biologia Vegetale, Università di Torino
mike.guether@unito.it

Abstract. The Stochastic Calculus of Wrapped Compartments (SCWC) is a recently proposed variant of the Stochastic Calculus of Looping Sequences (SCLS), a language for the representation and simulation of biological systems. In this work we apply SCWC to model a newly discovered ammonium transporter. This transporter is believed to play a fundamental role for plant mineral acquisition, which takes place in the *arbuscular mycorrhiza*, the most wide-spread plant-fungus symbiosis on earth. Investigating this kind of symbiosis is considered one of the most promising ways to develop methods to nurture plants in more natural manners, avoiding the complex chemical productions used nowadays to produce artificial fertilizers. In our experiments the passage of NH_3 / NH_4^+ from the fungus to the plant has been dissected in known and hypothetical mechanisms; with the model so far we have been able to simulate the behavior of the system under different conditions. Our simulations confirmed some of the latest experimental results about the LjAMT2;2 transporter. Moreover, by comparing the behaviour of LjAMT2;2 with the behaviour of another ammonium transporter which exists in plants, viz. LjAMT1;1, our simulations support an hypothesis about why LjAMT2;2 is so selectively expressed in arbusculated cells.

1 Introduction

Given the central role of agriculture in worldwide economy, several ways to optimize the use of costly artificial fertilizers are now being actively pursued. One approach is to find methods to nurture plants in more "natural" manners, avoiding the complex chemical production processes used today. In the last decade the Arbuscular Mycorrhiza (AM), the most widespread symbiosis between plants and fungi, got into the focus of research because of its potential as a natural plant

---

[*] This research is founded by the BioBITs Project (*Converging Technologies* 2007, area: Biotechnology-ICT), Regione Piemonte.

C. Priami et al. (Eds.): Trans. on Comput. Syst. Biol. XIII, LNBI 6575, pp. 85–109, 2011.

fertilizer. Briefly, fungi help plants to acquire nutrients as phosphorus (P) and nitrogen (N) from the soil whereas the plant supplies the fungus with energy in form of carbohydrates [34]. The exchange of these nutrients is supposed to occur mainly at the eponymous arbuscules, a specialized fungal structure formed inside the cells of the plant root. The arbuscules are characterized by a juxtaposition of a fungal and a plant cell membrane where a very active interchange of nutrients is facilitated by several membrane transporters. These transporters are surface proteins that facilitate membrane crossing of molecules which, because of their inherent chemical nature, are not freely diffusible.

Since almost all cells in the majority of multicellular organisms share the same genome, modern theories point out that morphological and functional differences between them are mainly driven by different genes expression [2]. Thanks to the latest experimental novelties [41,32] a precise analysis of which genes are expressed in a single tissue is attainable; therefore it is possible to identify genes that are pivotal in specific compartments and then study their biological function. Following this route a new membrane transporter has been discovered by expression analysis and further characterized [20]. This transporter is situated on the plant cell membrane which is directly opposite to the fungal membrane, located in the arbuscules. Various experimental evidence points out that this transporter binds to an NH_4^+ moiety outside the plant cell, deprotonates it, and mediates inner transfer of NH_3, which is then used as a nitrogen source, leaving an H^+ ion outside. The AM symbiosis is far from being unraveled: the majority of fungal transporters and many of the chemical gradients and energetic drives of the symbiotic interchanges are unknown. Therefore, a valuable task would be to model *in silico* these conditions and run simulations against the experimental evidence available so far about this transporter. Conceivably, this approach will provide biologists with working hypotheses and conceptual frameworks for future biological validation.

In computer science, several modelling languages for the representation and simulation of biological systems behaviour have been proposed. Automata-based models [4,30] have the advantage of allowing the direct use of many verification tools such as model checkers. Rewrite systems [13,37,9,1] usually allow describing biological systems with a notation that can be easily understood by biologists. Compositionality allows studying the behaviour of a system componentwise. Both automata-like models and rewrite systems present, in general, problems from the point of view of compositionality, which, instead, is in general ensured by process calculi, included those commonly used to describe biological systems [38,36,10]. The Stochastic Calculus of Looping Sequences (SCLS) [7] (see also [9,8,31]) combines the simplicity of notation of rewrite systems with the advantage of a form of compositionality. It also allows a rather simple and accurate description of biological membranes and their interactions with the environment.

The Stochastic Calculus of Wrapped Compartments (SCWC) [12] is a variant of SCLS. It has been designed with the aim of strongly simplifying the development of efficient automatic tools for the analysis of biological systems, while keeping the same expressiveness. The main simplification consists in the removal

of the sequencing operator, thus lightening the formal treatment of the patterns
to be matched in a term (whose complexity in SCLS is strongly affected by the
variables matching in the sequences).

In this work we apply SCWC to model the mentioned transporter. This
transporter is differently expressed in arbuscular cells and is believed to play a
fundamental role in the nutrients uptake which takes place in the context of
plants-fungi symbiosis at the root level. This symbiosis is one of the main focuses of the BioBITs project, due to its relevant role in agriculture.

On these premises, the aim of this work is to model the interchange between
the fungus-plant interface and the plant cells (using the NH_3/NH_4^+ turnover
as a reference system) and to support an hypothesis about why LjAMT2;2 is
so selectively expressed in arbusculated cells (by comparing its behaviour with
the behaviour of the LjAMT1;1 transporter, which exists in plants but is not
expressed in arbusculated cells). This could disclose which is the driving power
of the net nitrogen flux inside plant cells (still unknown at the chemical level).
Furthermore, this information may also be exploited to model other, and so far
poorly characterized, transporters.

Outline. Section 2 introduces the syntax and the semantics of the SCWC
and shows some modelling guidelines. Section 3 presents the SCWC representations of the ammonium transporter and our experimental results, discussed in
Section 4. Section 5 concludes by outlining some further work.

2   The Stochastic Calculus of Wrapped Compartments

In this section we briefly describe the Stochastic Calculus of Wrapped Compartments (SCWC) [12]. Like most modeling languages based on term rewriting
(notably [7]) an SCWC (biological) model consists of a term, representing the
system and a set of rewrite rules which models the transformations which determine its evolution.

Terms and Structural Congruence. A *term* of the SCWC calculus is intended to represent a biological system. A *term* is a sequence of *simple terms*.
Simple terms, ranged over by t, u, v, w, are built by means of the *compartment*
constructor, $(- \rfloor -)$, from a set $\mathcal{A}$ of *atomic elements* (*atoms* for short), ranged
over by a, b, c, d. The syntax of simple terms is given at the top of Figure 1. We
write $\bar{t}$ to denote a (possibly empty) sequence of simple terms $t_1 \cdots t_n$. Similarly,
with $\bar{a}$ we denote a (possibly empty) sequence of atoms. The set of simple terms
will be denoted by $\mathcal{T}$. The set of terms (sequences of simple terms) and the set of
sequences of atoms will be denoted by $\overline{\mathcal{T}}$ and $\overline{\mathcal{A}}$, respectively. Note that $\overline{\mathcal{A}} \subseteq \overline{\mathcal{T}}$.

A term $\bar{t} = t_1 \cdots t_n$ should be understood as the multiset containing the
simple terms $t_1, \ldots, t_n$. Therefore, we introduce a relation of structural congruence, following a standard approach in process algebra. The SCWC *structural
congruence* is the least equivalence relation on terms satisfying the rules given
at the bottom of Figure 1. From now on we will always consider terms modulo

Simple terms syntax

$$t \ ::= \ a \ \mid \ (\overline{a} \rfloor \overline{t})$$

Structural congruence

$$\overline{t} \ u \ w \ \overline{v} \equiv \overline{t} \ w \ u \ \overline{v} \qquad\qquad \text{if} \ \ \overline{a} \equiv \overline{b} \ \ \text{and} \ \ \overline{t} \equiv \overline{u} \ \ \text{then} \ \ (\overline{a} \rfloor \overline{t}) \equiv (\overline{b} \rfloor \overline{u})$$

Fig. 1. CWC term syntax and structural congruence rules

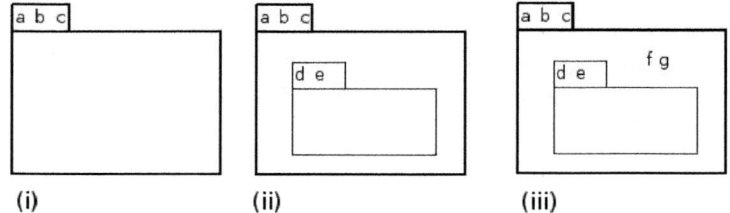

(i)　　　　　　　　**(ii)**　　　　　　　　**(iii)**

Fig. 2. (i) represents $(a \ b \ c \rfloor \bullet)$; (ii) represents $(a \ b \ c \rfloor (d \ e \rfloor \bullet))$; (iii) represents $(a \ b \ c \rfloor (d \ e \rfloor \bullet) \ f \ g)$

structural congruence.[1] Then a simple term is either an atom or a compartment $(\overline{a} \rfloor \overline{t})$ consisting of a *wrap* (represented by the multiset of atoms $\overline{a}$) and a *content* (represented by the term $\overline{t}$). We write the empty multiset as $\bullet$ and denote the union of two multisets $\overline{u}$ and $\overline{v}$ as $\overline{u} \ \overline{v}$. Let's extend the notion of a subset (denoted as usual as $\subseteq$) between terms interpreted as multisets.

An example of term is $a \ b \ (c \ d \rfloor e \ f)$ representing a multiset consisting of two atoms a and b (for instance two molecules) and a compartment $(c \ d \rfloor e \ f)$ which, in turn, consists of a wrap (a membrane) with two atoms c and d (for instance, two proteins) on its surface, and containing the atoms e (for instance, a molecule) and f (for instance a DNA strand). See Figure 2 for some graphical representations.

Rewrite Rules, Variables, Open Terms and Patterns. A rewrite rule is defined as a pair of terms (possibly containing variables), which represent the patterns defining the system transformations, together with a rate representing the speed of the modelled reaction. Rules are applicable to all subterms, identified by the notion of reduction context introduced below, which match the left-hand side of the rule via a proper instantiation of its variables. The system

---

[1] In the implementation of the derived tools it will be useful to consider a normalized representation of these terms suitable for efficient manipulation. In the description of the calculus given here however we will ignore implementation issues.

transformation is obtained by replacing the reduced subterm by the corresponding instance of the right-hand side of the rule.

In order to formally define the rewriting semantics, we introduce the notion of open term (a term containing variables) and pattern (an open term that may be used as left part of a rewrite rule). In order to respect the syntax of terms, we distinguish between "wrap variables" which may occur only in compartment wraps (and can be replaced only by multisets of atoms) and "term variables" which may only occur in compartment contents or at top level (and can be replaced by arbitrary terms). Therefore, we assume a set of *term variables*, $V_{\mathcal{T}}$, ranged over by X, Y, Z, and a set of *wrap variables*, $V_{\overline{\mathcal{A}}}$, ranged over by x, y, z. These two sets are disjoint. We denote by V the set of all variables $V_{\mathcal{T}} \cup V_{\overline{\mathcal{A}}}$, and with ρ any variable in V.

(i) *Open terms* are terms which may contain occurrences of wrap variables in compartment wraps and term variables in compartment contents or at top level. They can be seen as multisets of *simple open terms*. More formally, open terms, ranged over by O and simple open terms, ranged over by o, are defined in the following way:

$$O ::= \overline{o}$$
$$o ::= a \quad | \quad X \quad | \quad (\overline{a}\ \overline{x}\,\rfloor\,\overline{o})$$

(ii) An open term is *linear* if each variable occurs at most once in it.

(iii) *Simple patterns*, ranged over by S, are the linear open terms defined in the following way:

$$S \quad ::= \quad t \quad | \quad (\overline{a}\ x\,\rfloor\,\overline{S}\ X)$$

where t is an element of $\mathcal{T}$, $\overline{a}$ is an element of $\overline{\mathcal{A}}$, x is a variable in $V_{\overline{\mathcal{A}}}$, $\overline{S}$ is a possibly empty multiset of simple patterns and X is a variable in $V_{\mathcal{T}}$. We denote with $\mathcal{S}$ the set of simple patterns.

(iv) *Patterns*, ranged over by P, are the linear open terms defined in the following way:

$$P \quad ::= \quad S\ \overline{S}\ X$$

where $S\ \overline{S}$ is a nonempty multiset of simple patterns and X is an element of $V_{\mathcal{T}}$. We denote with $\mathcal{P}$ the set of patterns.

An *instantiation* is a partial function $\sigma : V \to \overline{\mathcal{T}}$. An instantiation must preserve the type of variables, thus for $X \in V_{\mathcal{T}}$ and $x \in V_{\overline{\mathcal{A}}}$ we have $\sigma(X) \in \overline{\mathcal{T}}$ and $\sigma(x) \in \overline{\mathcal{A}}$, respectively. Given $O \in \mathcal{O}$, with $O\sigma$ we denote the term obtained by replacing each occurrence of each variable $\rho \in V$ appearing in O with the corresponding term $\sigma(\rho)$.

Let Σ denote the set of all the possible instantiations and $Var(O)$ denote the set of variables appearing in $O \in \mathcal{O}$. A *rewrite rule* is a pair (P, O), denoted with $P \longmapsto O$, where $P \in \mathcal{P}$ and $O \in \mathcal{O}$ are such that $Var(O) \subseteq Var(P)$. A rewrite rule $P \longmapsto O$ then states that a subterm $P\sigma$, obtained by instantiating variables in P by some instantiation function σ, can be transformed into the subterm $O\sigma$.

We extend the notion of structural congruence to open terms in the natural way.

Contexts. The definition of reduction for SCWC systems is completed by resorting to the notion of reduction contexts. To this aim, as usual, the syntax of terms is enriched with a new element $\Box$ representing a hole. *Reduction context* (ranged over by C) are defined by:

$$C \ ::= \ \Box \ \mid \ (\bar{a} \rfloor C) \ \bar{t}$$

where $\bar{a} \in \overline{\mathcal{A}}$ and $\bar{t} \in \overline{\mathcal{T}}$. We denote with $\mathcal{C}$ the infinite set of contexts.

By definition, every context contains a single hole $\Box$. Let us assume $C, C' \in \mathcal{C}$. With $C[\bar{t}]$ we denote the term obtained by replacing $\Box$ with $\bar{t}$ in C; with $C[C']$ we denote context composition, whose result is the context obtained by replacing $\Box$ with C' in C. For example, given $C = (a\ b \rfloor \Box)\ i\ l$, $C' = (c\ d \rfloor \Box)\ g\ h$ and $\bar{t} = e\ f$, we get $C[C'[\bar{t}]] = (a\ b \rfloor (c\ d \rfloor e\ f)\ g\ h)\ i\ l$. Note that context holes take the place either of the whole term or of the whole content of a compartment. This allows to make contexts unambiguous in the following sense: For any term $\bar{t}$ if the term $\bar{u}$ occurs in $\bar{t}$ within a compartment content or at top level, then there are, modulo $\equiv$, a unique context C and a unique term $\bar{t}'$ such that $\bar{t} = C[\bar{t}']$ and $\bar{u} \subseteq \bar{t}'$.

A transformation of term $C[\bar{t}]$ determined by the application of a rewrite rule (P, O) can then be described in the following way:

$$\frac{P \longmapsto O \quad \sigma \in \Sigma \quad \bar{t} = P\sigma \quad \bar{u} = O\sigma \quad C \in \mathcal{C}}{C[\bar{t}] \longmapsto C[\bar{u}]}$$

Stochastic Reduction Semantics. The operational semantics of SCWC is defined by incorporating a collision-based stochastic framework along the lines of the one presented by Gillespie in [15], which is, *de facto*, the standard way to model quantitative aspects of biological systems. The idea of Gillespie's algorithm is that a rate constant is associated with each considered chemical reaction. Such a constant is obtained by multiplying the kinetic constant of the reaction by the number of possible combinations of reactants that may occur in the system. The resulting rate is then used as the parameter of an exponential distribution modelling the time spent between two occurrences of the considered chemical reaction. Following the law of mass action, it is necessary to count the number of reactants that are present in a system in order to compute the exact rate of a reaction. The same approach has been applied, for instance, to the stochastic π-calculus [35,36].

The use of exponential distributions to represent the (stochastic) time spent between two occurrences of chemical reactions allows describing the system as a *Continuous Time Markov Chain* (CTMC), and consequently allows verifying properties of the described system analytically and by means of stochastic model checkers.

We then associated to the rewriting rules of SCWC the kinetic constant k of the modeled chemical reaction. A *stochastic rewrite rule* is then a triple (P, O, k), denoted with $P \xmapsto{k} O$, where (P, O) is a rewrite rule and $k \in \mathbb{R}^{\geq 0}$.

The number of reactants in a reaction represented by a rewrite rule is evaluated considering the number of distinct occurrences, in the same context , of subterms to which the rule can be applied producing the same term . For instance in evaluating the application rate of the stochastic rewrite rule $R = a\ b \overset{k}{\longmapsto} c$ to the term $\bar{t} = a\ a\ b\ b$ we must consider the number of the possible combinations of reactants of the form $a\ b$ in $\bar{t}$. Since each occurrence of a can react with each occurrence of b, this number is 4. So the application rate of R is $k \cdot 4$.

The evaluation of the application rate of a reduction rule containing variables is more complicated since there can be many different ways in which variables can be instantiated to match the subterm to be reduced, and this must be considered to correctly evaluate the application rate. Given two terms $\bar{t}, \bar{u}$ and a reduction rule R, we can compute the number of possible applications of the rule R to the term $\bar{t}$ resulting in the term $\bar{u}$. We denote this number by $\mathcal{O}(R, \bar{t}, \bar{u})$. We refer to [12] for more details and explanation.[2] In particular, the function $\mathcal{O}$ is analogous to the one defined for SCLS in [7].

A SCWC *system* over a set $\mathcal{A}$ of atoms is represented by a set $\mathcal{F}_{\mathcal{A}}$ ($\mathcal{F}$ for short when $\mathcal{A}$ is understood) of stochastic rewrite rules over $\mathcal{A}$. Given an SCWC system, the *reduction semantics* of SCWC is the least labelled transition relation satisfying the following rule:

$$\frac{R = P \overset{k}{\longmapsto} O \in \mathcal{R} \qquad \sigma \in \Sigma \qquad \bar{t} = P\sigma \qquad \bar{u} = O\sigma \qquad C \in \mathcal{C}}{C[\bar{t}] \overset{k \cdot \mathcal{O}(R, \bar{t}, \bar{u})}{\longmapsto} C[\bar{u}]}$$

The rate of the reduction is then obtained as the product of the rewrite rate constant and the number of occurrences of the rule within the starting term (thus counting the exact number of reactants to which the rule can be applied and which produce the same result). The rate associated with each transition in the stochastic reduction semantics is the parameter of an exponential distribution that characterizes the stochastic behaviour of the activity corresponding to the applied rewrite rule. The stochastic semantics is essentially a CTMC. A standard simulation procedure that corresponds to Gillespie's simulation algorithm [15] can be followed. A prototype implementation of SCWC, based on Gillespie's algorithm, is available [3].

In the following we will write:

(i) $S\ \overline{S} \overset{k}{\Rightarrow} O$ as short for $S\ \overline{S}\ X \overset{k}{\longmapsto} O\ X$, where the variable X does not occur in $Var(S\ \overline{S}) \cup Var(O)$.

(ii) $a\ \overline{a} \overset{k}{\longrightarrow} \overline{b}$ as short for the rewrite rule $(a\ \overline{a}\ x \rfloor Y)\ Z \overset{k}{\longmapsto} (\overline{b}\ x \rfloor Y)\ Z$, where the variables x, Y, Z are distinct.

---

[2] The semantics of the calculus described in [12] is slightly more general. In particular, rules are defined with rate functions (instead of constant rates), which model the speed of a reaction in a parametric way. For the application investigated in this paper it is enough to use constant kinetics, therefore we presented here a simplified semantics.

Table 1. Guidelines for the abstraction of biomolecular events into SCWC

Biomolecular Event	Examples of SCWC Rewrite Rules
State change	$a \Rightarrow b$
Complexation	$a \; b \Rightarrow c$
Decomplexation	$c \Rightarrow a \; b$
Catalysis	$c \; S \; \overline{S} \Rightarrow c \; O$ where $S \; \overline{S} \Rightarrow O$ is the catalyzed event
State change on membrane	$a \rightarrow b$
Complexation	$a \; b \rightarrow c$
on membrane	$a \; (b \; x \rfloor X) \Rightarrow (c \; x \rfloor X)$
	$(b \; x \rfloor a \; X) \Rightarrow (c \; x \rfloor X)$
Decomplexation	$c \rightarrow a \; b$
on membrane	$(c \; x \rfloor X) \Rightarrow a \; (b \; x \rfloor X)$
	$(c \; x \rfloor X) \Rightarrow (b \; x \rfloor a \; X)$
Catalysis on membrane	$c \; a \; \overline{a} \rightarrow c \; \overline{b}$ where $a \; \overline{a} \rightarrow \overline{b}$ is the catalyzed event
Membrane crossing	$a \; (x \rfloor X) \Rightarrow (x \rfloor a \; X)$
	$(x \rfloor a \; X) \Rightarrow a \; (x \rfloor X)$
Catalyzed	$a \; (b \; x \rfloor X) \Rightarrow (b \; x \rfloor a \; X)$
membrane crossing	$(b \; x \rfloor a \; X) \Rightarrow a \; (b \; x \rfloor X)$
Membrane joining	$a \; (x \rfloor X) \Rightarrow (a \; x \rfloor X)$
	$(x \rfloor a \; X) \Rightarrow (a \; x \rfloor X)$
Catalyzed	$a \; (b \; x \rfloor X) \Rightarrow (a \; b \; x \rfloor X)$
membrane joining	$(b \; x \rfloor a \; X) \Rightarrow (a \; b \; x \rfloor X)$
	$(x \rfloor a \; b \; X) \Rightarrow (a \; x \rfloor b \; X)$
Membrane fusion	$(a \; x \rfloor X) \; (b \; y \rfloor Y) \Rightarrow (a \; x \; b \; y \rfloor X \; Y)$
Vesicle dynamics	$(x \rfloor X \; (a \; y \rfloor Y)) \Rightarrow Y \; (a \; y \; x \rfloor X)$

Modelling Guidelines. In this section we will give some explanations and general hints about how SCWC could be used to represent the behaviour of various biological systems.

In rewrite systems, such as SCWC, entities are usually represented by terms of the rewrite system, and events by rewrite rules. First of all, we should select the biomolecular entities of interest. Since we want to describe cells, we consider molecular populations and membranes. Molecular populations are groups of molecules that are in the same compartment of the cells and inside them. As we have said before, molecules can be of many types: we classify them as proteins, chemical moieties and other molecules.

Membranes are considered as elementary objects: we do not describe them at the level of the phospholipids they are made of. The only interesting property of a membrane are that it may have a content (hence, create a compartment) and that in its phospholipid bilayer various proteins are embedded, which act for example as transporters and receptors. Since membranes are represented as multisets of the embedded structures, we are modelling a fluid mosaic in which the membranes become similar to a two-dimensional liquid where molecules can diffuse more or less freely [43].

Now, we select the biomolecular events of interest. The simplest kind of event is the change of state of an elementary object. Then, there are interactions between molecules: in particular complexation, decomplexation and catalysis. Interactions could take place between simple molecules, depicted as single symbols, or between membranes and molecules: for example a molecule may cross or join a membrane. Finally, there are also interactions between membranes: in this case there may be many kinds of interactions (fusion, vesicle dynamics, etc. . .).

Table 1 lists the guidelines (taken from [12]) for the abstraction into SCWC rules of the biomolecular events we will use in our application.[3] Entities are associated with SCWC terms: elementary objects (genes, domains, etc...) are modelled as multisets of atoms, molecular populations as SCWC terms, and membranes as atom multisets. Biomolecular events are associated with SCWC rewrite rules.

3   Modelling the Ammonium Transporter with SCWC

The scheme in Figure 3 (taken from [20]) illustrates nitrogen, phosphorus and carbohydrate exchanges at the mycorrhizal interface according to previous works and the results of [20]. **(a1-2)** NH_3/NH_4^+ is released in the arbuscules from arginine which is transported from the extra- to the intraradical fungal structures [18]. NH_3/NH_4^+ is released by so far unknown mechanisms (transporter, diffusion **(a1)** or vesicle-mediated **(a2)**) into the periarbuscular space (PAS) where, due to the acidic environment, its ratio shifts towards NH_4^+ ($> 99.99\%$). **(b)** The acidity of the interfacial apoplast is established by plant and fungal H^+-ATPases [23,6] thus providing the energy for H^+-dependent transport processes. **(c)** The NH_4^+ ion is deprotonated prior to its transport across the plant membrane via the LjAMT2;2 protein and released in its uncharged NH_3 form into the plant cytoplasm. The NH_3/NH_4^+ acquired by the plant is either transported into adjacent cells or immediately incorporated into amino acids (AA). **(d)** Phosphate is released by so far unknown transporters into the interfacial apoplast. **(e)** The uptake of phosphate on the plant side then is mediated by mycorrhiza-specific Pi-transporters [27,20].[4] **(f)** AM fungi might control the net Pi-release by their own Pi-transporters which may reacquire phosphate from the periarbuscular space [6]. **(g)** Plant derived carbon is released into the PAS probably as sucrose and then cleaved into hexoses by sucrose synthases [24] or invertases [40]. AM fungi then acquire hexoses [42,44] and transport them over their membrane by so far unknown hexose transporters. It is likely that these transporters are proton co-transporter as the GpMST1 described for the glomeromycotan fungus Geosiphon pyriformis [33]. Exchange of nutrients between

---

[3] Kinetics are omitted from the table for simplicity.
[4] Where Pi stands for inorganic phosphate.

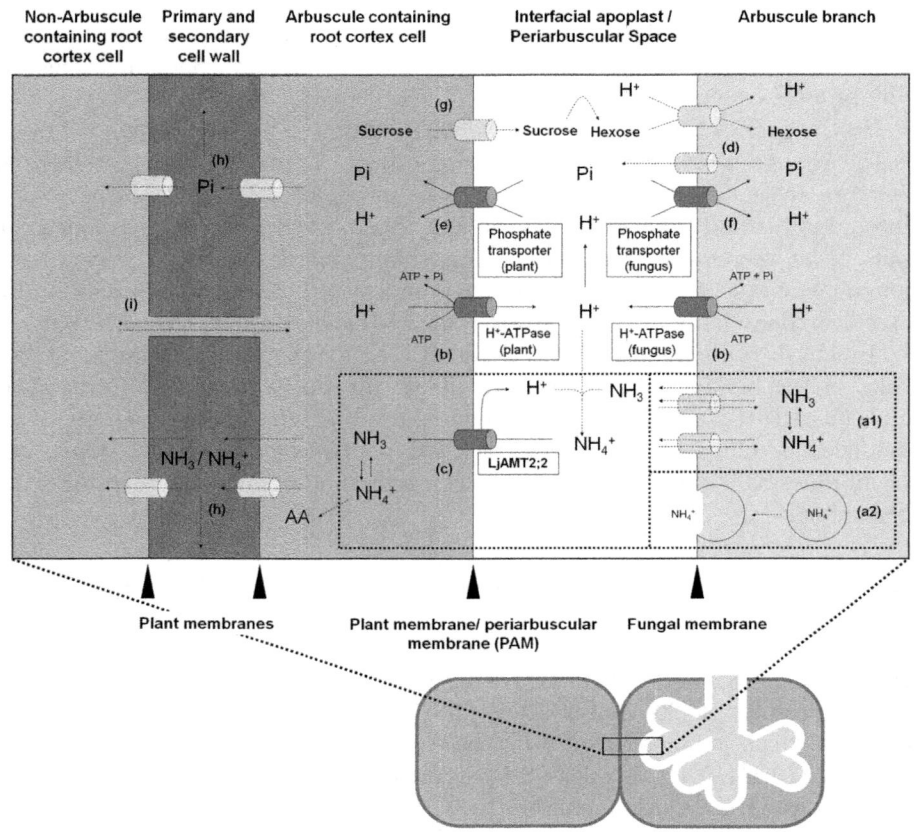

Fig. 3. Nitrogen, phosphorus and carbohydrate exchanges at the mycorrhizal interface

arbusculated cells and non-colonized cortical cells can occur by apoplastic (**h**) or symplastic (**i**) ways.

In this paper we focus our investigation on the sectors labelled with (**c**), (**a1**) and (**a2**). Namely, we will present SCWC models for the equilibrium between NH_4^+ and NH_3 and the uptake by the LjAMT2;2 transporter (**c**), and the exchange of NH_4^+ from the fungus to the interspatial level (**a1-2**). We will also analyze LjAMT2;2 role in the AM symbiosis by comparing it with another known ammonium transporter, LjAMT1;1. The choice of SCWC is motivated by the fact that membranes, membrane elements (like LjAMT2;2) and the involved reactions can be represented in it in a quite natural way.

The simulations illustrated in this section are done with the SCWC prototype simulator [3]. In the following we will use a more compact notation to represent multisets of the same atom, namely, we will write $a \times n$ to denote the multiset of n atomic elements a. Moreover, to simplify the counting mechanism, we might use different names for the same molecule when it belongs to different compartments.

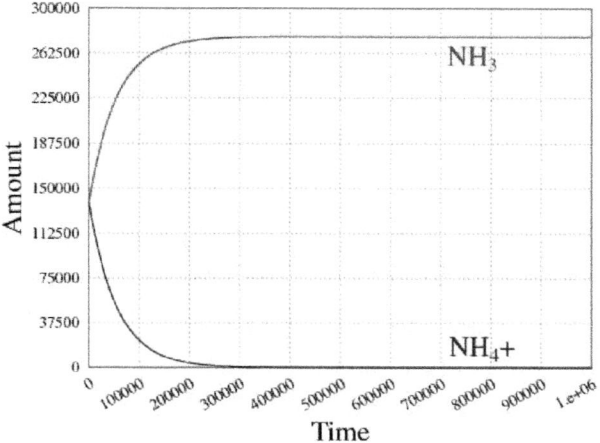

Fig. 4. Extracellular equilibrium between NH_3 and NH_4^+

For instance, occurrences of the molecule NH_3 inside the plant cell are referred to as NH_3_inside.

3.1 NH_3/NH_4^+ Equilibrium

We decided to start modelling a simplified pH equilibrium, at the interspatial level (right part of section (**c**) in Figure 3), without considering H_2O, H^+ and OH^-; therefore we tuned the reaction rates in order to reach the correct percentages of NH_3 over total NH_3/NH_4^+ in the different compartments. Like these all the rates and initial terms used in this work are obtained by manual adjustments made looking at the simulations results and trying to keep simulations times acceptable - we plan to refine these rates and numbers in future work to reflect biological data when they become available. Following [20], we consider an extracellular pH of 4.5 [21]. In such conditions, the percentage of molecules of NH_3 over the sum $NH_3 + NH_4^+$ should be around 0.002. The reaction we considered is the following:

$$NH_3 \underset{k_2}{\overset{k_1}{\rightleftharpoons}} NH_4^+$$

with $k_1 = 0.018 \times 10^{-3}$ and $k_2 = 0.562 \times 10^{-9}$. One can translate this reaction with the SCWC rules.

$$NH_3 \overset{k_1}{\Rightarrow} NH_4^+ \tag{R1}$$

$$NH_4^+ \overset{k_2}{\Rightarrow} NH_3 \tag{R2}$$

In Figure 4 we show the results of this first simulation given the initial term $\bar{t} = NH_3 \times 138238\ NH_4^+ \times 138238$.

This equilibrium is different at the intracellular level (pH around 7 and 8) [17], so we use two new rules to model the transformations of NH_3 and NH_4^+ inside the cell, namely:

$$NH_3\text{\_inside} \overset{k_1}{\Rightarrow} NH_4^+\text{\_inside} \tag{R3}$$

$$NH_4^+\text{\_inside} \overset{k_2'}{\Rightarrow} NH_3\text{\_inside} \tag{R4}$$

where $k_2' = 0.562 \times 10^{-6}$.

3.2 LjAMT2;2 Uptake

We can now present the SCWC model of the uptake of the LjAMT2;2 transporter (left part of section (c) in Figure 3). We add a compartment modelling an arbusculated plant cell. Since we are only interested in the work done by the LjAMT2;2 transporter, we consider a membrane containing this single element. The work of the transporter is modelled by the rule:

$$NH_4^+ \; (LjAMT2 \; x \rfloor X) \overset{k_t}{\Rightarrow} H^+ \; (LjAMT2 \; x \rfloor X \; NH_3\text{\_inside}) \tag{R5}$$

where $k_t = 0.1 \times 10^{-5}$.

We can investigate the uptake rate of the transporter at different initial concentrations of NH_3 and NH_4^+. Figure 5 and Figure 6 show the results for the initial terms:

$$\bar{t} = NH_3 \times 776 \; NH_4^+ \times 276400 \; (LjAMT2 \rfloor \bullet)$$
$$\bar{u} = NH_3 \times 276400 \; NH_4^+ \times 776 \; (LjAMT2 \rfloor \bullet)$$

where the graphs above represent the whole simulations, while the ones below are a magnification of their initial segment.

We can also investigate the uptake rate of the transporter at different extracellular pH. Namely, we consider an extracellular pH equal to the intracellular one (pH around 7 and 8), obtained by imposing $R1$ and $R2$ equal to $R3$ and $R4$, respectively, i.e. $k_2 = k_2'$. Figure 7 shows the results for the initial term $\bar{v} = NH_3 \times 138238 \; NH_4^+ \times 138238 \; (LjAMT2 \rfloor \bullet)$.

Since now we modeled the transporter supposing that no active form of energy is required to do the actual work - which means that the NH_4^+ gradient between the cell and the extracellular ambient is sufficient to determine a net uptake. The predicted tridimensional structure of LjAMT2;2 suggests that it does not use ATP [5] as an energy source [20], nevertheless trying to model an "energy consumption" scenario is interesting to make some comparisons. Since this is only a proof of concept there is no need to specify here in which form this energy is going to be provided. Furthermore, as long as we are only interested in comparing the initial rates of uptake, we can avoid defining rules that regenerate energy in the cell. Therefore, rule R5 modelling the transporter role can be modified as follows:

$$NH_4^+ \; (LjAMT2 \; x \rfloor ENERGY \; X) \overset{k_t'}{\Rightarrow} \tag{R5'}$$
$$H^+ \; (LjAMT2 \; x \rfloor NH_3\text{\_inside} \; X)$$

---

[5] ATP is the "molecular unit of currency" of intracellular energy transfer [28] and is used by many transporters that work against chemical gradients.

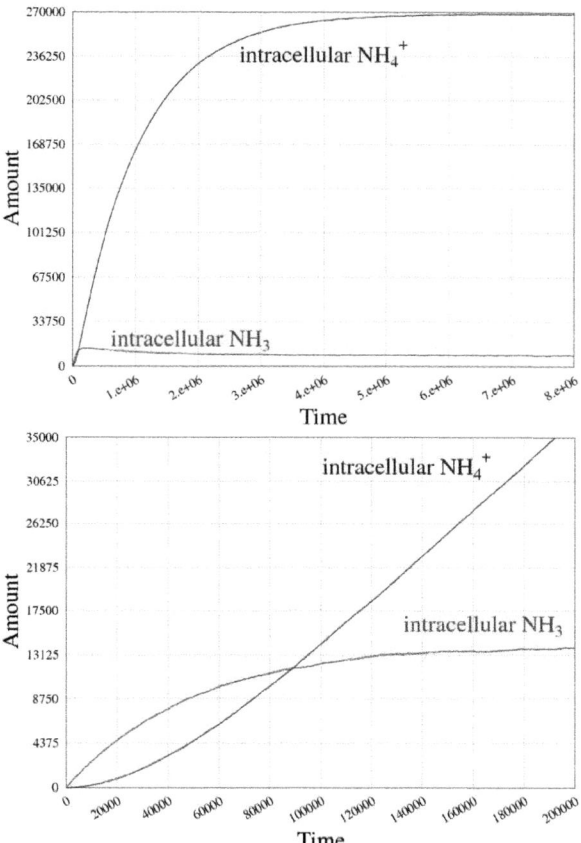

Fig. 5. At high NH_4^+ concentration

which consumes an element of energy within the cell. We also make this reaction slower, since it is now catalysed by the concentration of the *ENERGY*
element, actually, we set $k'_t = 0.1 \times 10^{-10}$. Given the initial term $\overline{w} = NH_3 \times$
138238 $NH_4^+ \times 138238$ $(LjAMT2 \rfloor ENERGY \times 100000)$ we obtain the simulation result in Figure 8. Note that the uptake work of the transporter terminates
when the *ENERGY* inside the cell is completely exhausted.

3.3 NH_4^+ Diffusing from the Fungus

We now model the diffusion of NH_4^+ from the fungus to the extracellular level
(sections **(a1)**, and **(a2)** of Figure 3). In section **(a1)** of the figure, the passage of
NH_4^+ to the interfacial periarbuscular space happens by diffusion. We can model
this phenomenon by adding a new compartment, representing the fungus, from

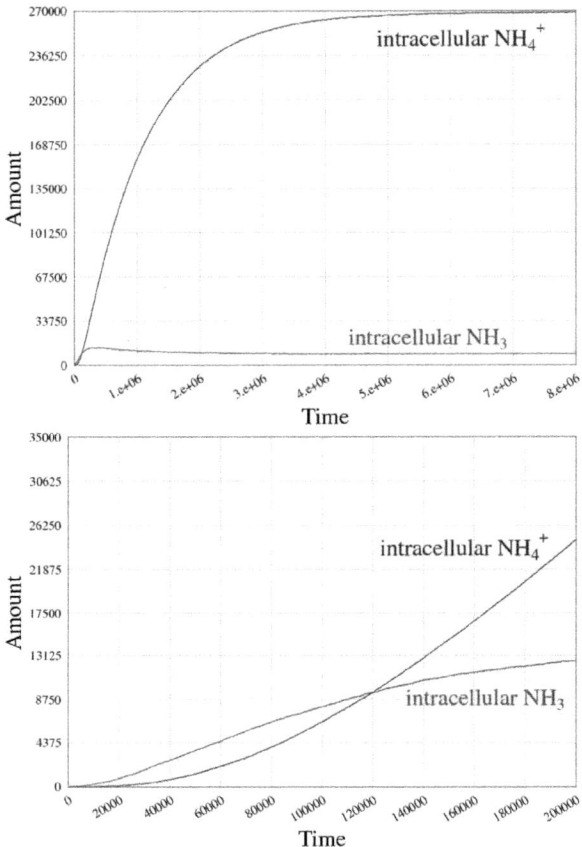

Fig. 6. At low NH_4^+ concentration

which NH_4^+ flows towards the fungus-plant interface. This could be modelled through the rule:

$$(FungMembr \; x \rfloor NH_4^+ \; X) \stackrel{k_f}{\Rightarrow} NH_4^+ \; (FungMembr \; x \rfloor X) \qquad \text{(R6)}$$

By varying the value of the rate k_f one might model different externalization speeds and thus test different hypotheses about the underlying mechanism. In Figure 9 we give the simulation result, with three different magnification levels, going from the whole simulation to the initial parts, obtained from the initial term $\bar{t}_f = (FungMembr \; \rfloor \; NH_4^+ \times 2764677) \; (LjAMT2 \rfloor \bullet)$ with $k_f = 1$. In the initial part, one can see how fast, in this case, NH_4^+ diffuses into the extracellular space. In Figure 10 we give the simulation result obtained from the same initial term $\bar{t}_f$ with a slower diffusion rate, namely $k_f = 0.01 \times 10^{-3}$.

Additionally, we would like to remark, without going into the simulation details, how we can model in a rather natural way the portion **(a2)** of Figure 3

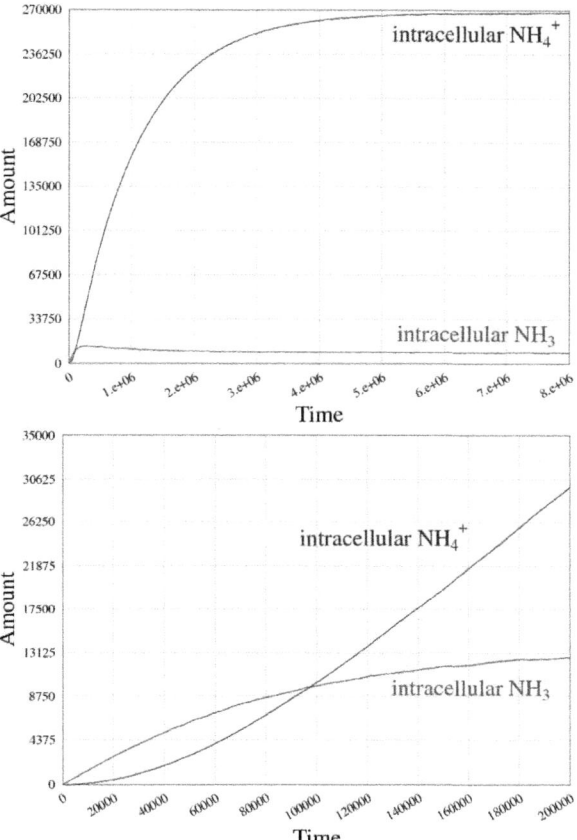

Fig. 7. At extracellular pH=7

in SCWC. Namely, we need some rules to produce vesicles containing NH_4^+ molecules within the fungal cell. Once the vesicle is formed, another rule drives its exocytosis towards the interfacial space, and thus the diffusion of the previously encapsulated NH_4^+ molecules. The necessary rules are given in the following:

$$(FungMembr \; x \mid X) \stackrel{k_{\underline{c}}}{\Rightarrow} (FungMembr \; x \mid X \; (Vesicle \mid \bullet)) \qquad (R7)$$

$$(FungMembr \; x \mid NH_4^+ \; Y \; (Vesicle \; y \mid X)) \stackrel{k_{\underline{c}}}{\Rightarrow} \qquad (R8)$$
$$(FungMembr \; x \mid Y \; (Vesicle \; y \mid NH_4^+ \; X))$$

$$(FungMembr \; x \mid Y \; (Vesicle \; y \mid X)) \stackrel{k_{\underline{c}}}{\Rightarrow} X \; (FungMembr \; x \; y \mid Y) \qquad (R9)$$

Where rule R7 models the creation of a vesicle, rule R8 model the encapsulation of an NH_4^+ molecule within the vesicle and rule R9 models the exocytosis of the vesicle content.

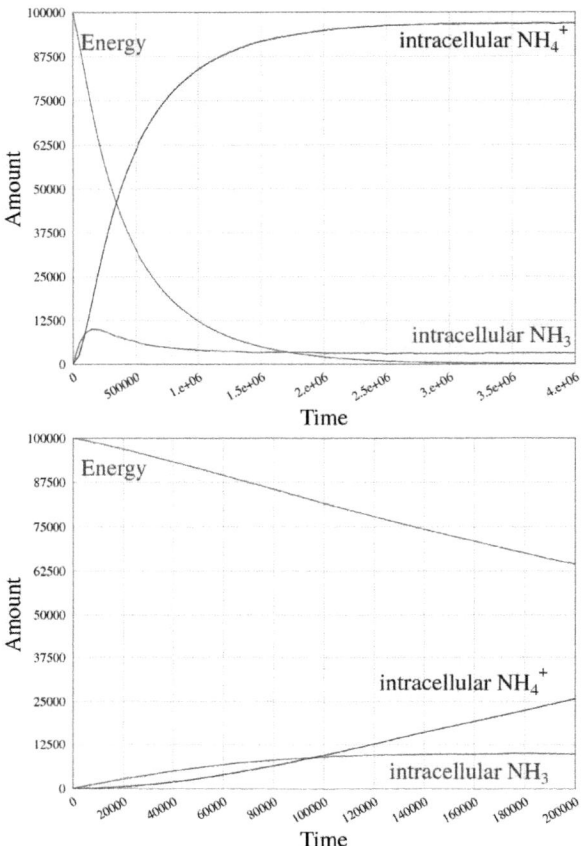

Fig. 8. LjAMT2;2 with active energy

3.4   Emplicit pH Representation

To be able to further investigate the delicate equilibria that are established in the AM symbiosis we needed a model that deals with H^+ and OH^- and therefore could offer more precise analysis opportunities on the system - to correctly model chemical reactions at different pHs it is important to find a set of rules that is capable of reaching and keeping the right ratio between H^+ and OH^-, even when used with other rules that comprises these ions, adding or removing them.

To avoid the need of huge quantities of water molecules only hydrogen ions and hydroxide are considered and not the process of water dissociation (as long as water is 55.5 M while, for example, at pH 4.5 hydrogen is 3.2×10^{-5} M and hydroxyl is 3.2×10^{-10} M and we have to use natural numbers to represent the quantities of molecules in the simulations it would be cumbersome to consider water).

Thus the rules simply have to "create" and "destroy" the ions: H^+ has to be destroyed considering its quantity and generated considering OH^- quantity and

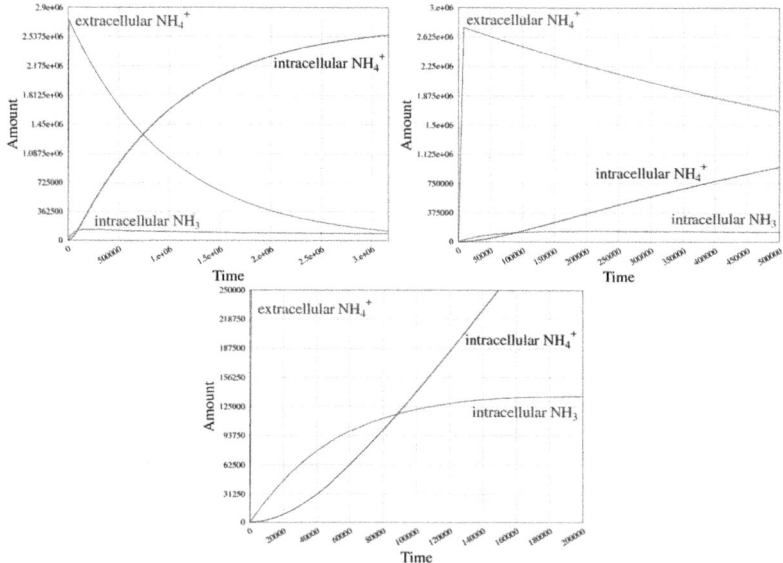

Fig. 9. Diffusing NH_4^+ from the fungus, $k_f = 1$

the same should be done for OH^-;[6] different pH will be obtained changing the rates of these rules.

The rules are easy to define: $H^+ \stackrel{k1}{\Rightarrow} \bullet$ and $OH^- \stackrel{k2}{\Rightarrow} \bullet$ for the rules that destroy ions and $H^+ \stackrel{k3}{\Rightarrow} H^+ \ OH^-$ and $OH^- \stackrel{k4}{\Rightarrow} OH^- \ H^+$ to create them, the stochastic simulation correctly applies them with an application rate that depends on the defined ks and the given ion numbers in the term to which they are applied. In this way two couples of rules are capable of maintaining a proper ratio between H^+ and OH^-: to obtain different pH it is enough to tune their rates in order to reflect the desiderate ratio.

We defined the correct rates for different pHs, for example pH 4.5, which is necessary to model the periarbuscular space and is characterized by a ratio between H^+ and OH^- of 10^5.

3.5 NH_3/NH_4^+ Equilibrium and LjAMT2;2

After the definition of proper rules for pH, we had to drive the exchange between ammonium cations and ammonia, with rules that should be capable of reaching and keeping the right ratio for these molecules at a given pH. Due to the explicit pH model the rules were changed with respect to the previous ones:

$$NH_3 \ H^+ \stackrel{k1}{\Rightarrow} NH_4^+ \tag{R1'}$$

---

[6] In such a way we abstract the reaction $H_2O \rightleftharpoons OH^- + H^+$ without considering explicitly the amount of water molecules.

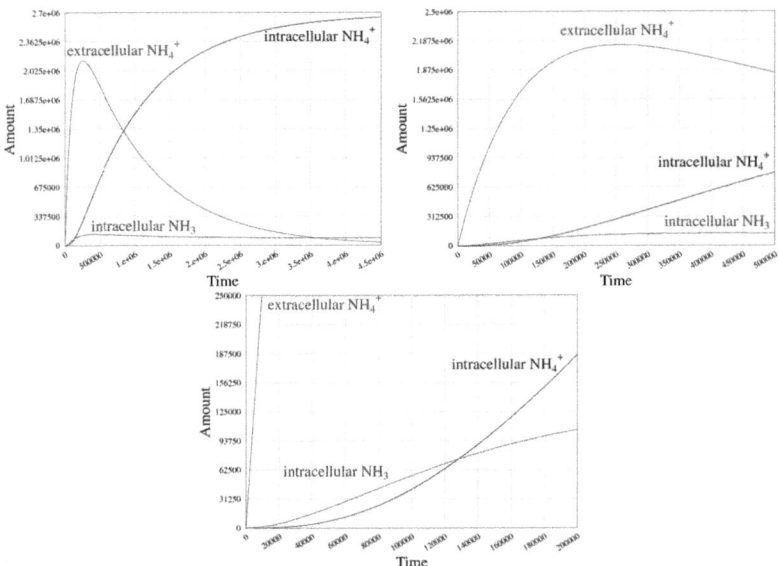

Fig. 10. Diffusing NH_4^+ from the fungus, $k_f = 0.01 * 10^{-3}$

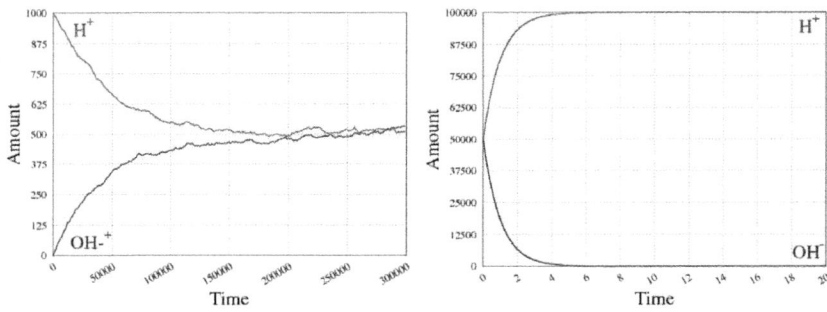

Fig. 11. pH 7 and 4.5 with correct ratios reached

$$NH_4^+ \overset{k2}{\Rightarrow} NH_3 \; H^+ \tag{R2'}$$

Rule (R1') represents ammonia which becomes protonated binding a free hydrogen ion, while rule (R2') is the converse reaction. We did not consider other reactions involving water, such as $NH_3 \; H_2O \overset{k3}{\Rightarrow} NH_4^+ \; OH^-$, for the previously explained reasons.

Several simulation were needed to tune the right rates, which were defined for pH 7, pH 4.5 and pH 8.6 - more extreme pHs are unlikely in our biological domain and they are difficult to model due to the ratios that have to be reached (1.7×10^9 between H^+ and OH^- at pH 2.6), which force to run simulations with huge molecule numbers.

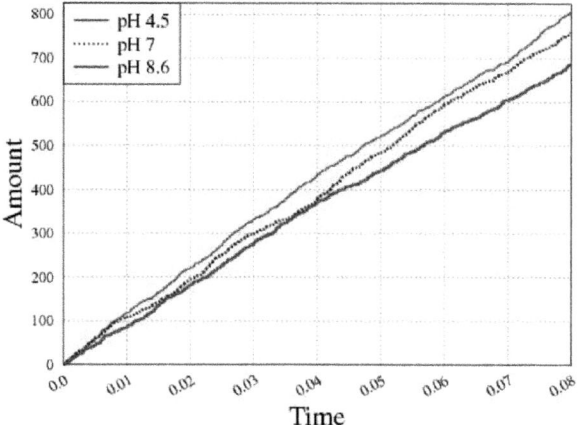

Fig. 12. LjAMT2;2 uptake (internalized NH3), comparison between different extracellular pHs

Having defined all these rules the next step was to add LjAMT2;2 and try to confirm some of the results obtained with the first model, for example the pH dependent uptake rate. Figure 12 shows a plot that represents the internalized NH_3 versus time with different periarbuscular pHs. The three simulations, started with "steady state" quantities of H^+ and OH^- and of NH_3 and NH_4^+ outside the cell (according to the chosen pH), while the cell started with no ammonia; they still have only a cell with a single transporter on the membrane - to be able to compare the internalization rate with sufficient numbers we changed the rate for the transport rule, namely $k_t = 0.1 \times 10^{-2}$.

3.6 Comparison with LjAMT1;1

To further investigate the role of LjAMT2;2 in the context of AM and its peculiar mechanism of transport, which does not depend on the H^+ gradient and seems otherwise to have a role in its maintenance (by expelling the H^+ gotten from the NH_4^+ molecule), we compared it with another ammonium transporter which exists in plants but is not so selectively expressed in arbusculated cells: LjAMT1;1 [39]; this is a transporter for NH_4^+ that does not expel H^+ in the periarbuscular space, therefore it internalizes directly an NH_4^+ molecule.

It is interesting to try to understand if LjAMT2;2 has a role which is synergic with other transporters in the AM symbiosis which relies on the H^+ gradient, such as those for the phosphates on the plant side or for the carbohydrates in the fungi and if other ammonia transporters, like LjAMT1;1, would instead "compete" with other transporters consuming hydrogen.

Modelling what would happen if in the arbusculated cell LjAMT1;1 will be the principal ammonia transporter, instead of LjAMT2;2, requires only to change

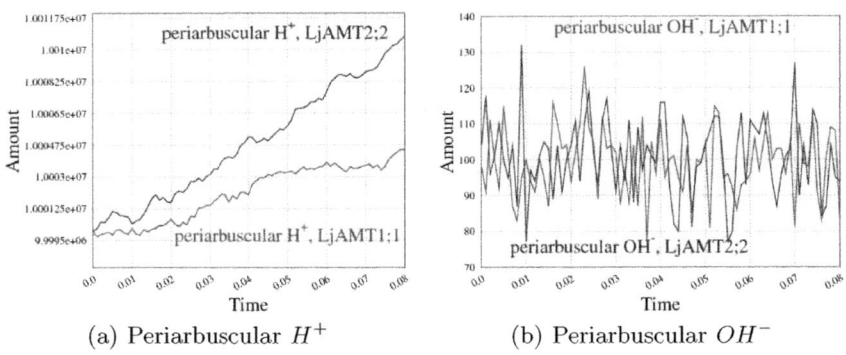

(a) Periarbuscular H^+ (b) Periarbuscular OH^-

Fig. 13. Comparison between LjAMT1;1 and LjAMT2;2

the transporter rule with respect to the simulation with the periarbuscular pH at 4.5 (which is the normal one) shown in the previous section:

$$NH_4^+ \ (LjAMT1 \ x \rfloor X) \overset{k_t}{\Rightarrow} (LjAMT1 \ x \rfloor X \ NH_4^+\_inside) \qquad \text{(R7')}$$

Note that the rate is the same for both rules.

Figure 13 represents the periarbuscular quantities of H^+ and OH^- for two simulations which had all the same rules except for the transporter one and started from the terms:

$$NH_4^+ \times 10^7 \ NH_3 \times 200 \ OH^- \times 100 \ H^+ \times 10^7 \ (LjAMT2 \rfloor \bullet)$$

$$NH_4^+ \times 10^7 \ NH_3 \times 200 \ OH^- \times 100 \ H^+ \times 10^7 \ (LjAMT1 \rfloor \bullet)$$

4   Discussion

We dissected the route for the passage of NH_3 / NH_4^+ from the fungus to the plant in known and hypothetical mechanisms which were transformed in rules. Further, also the properties of the different compartments and their influence on the transported molecules were included, thus giving a first model for the simulation of the nutrients transfer. With the model so far we can simulate the behaviour of the system when varying parameters as the different compartments pH, the initial substrate concentrations, the transport/diffusion speeds and the energy supply.

We can start comparing the two simulations with the plant cell with the LjAMT2;2 transporter placed in different extracellular situations: low NH_4^+ concentration (Figure 6) or high NH_4^+ concentration (Figure 5). As a natural consequence of the greater concentration, the ammonium uptake is faster when the simulation starts with more NH_4^+, as long as the LjAMT2;2 can readily import it. The real situation should be similar to this simulation, assuming that the level of extracellular NH_4^+ / NH_3 is stable, meaning an active symbiosis.

The simulation which represents an extracellular pH around 7 (Figure 7) shows a decreased internalisation speed with respect to the simulation in Figure 5, as could be inferred from the concentrations of NH4_inside and NH3_inside in the plots on the right (focusing on the initial activity): this supports experimental data about the pH-dependent activity of the transporter and suggests that the extracellular pH is fundamental to achieve a sufficient ammonium uptake for the plants. It is noticeable how the initial uptake rate in this case is higher, despite the neutral pH, than the rate obtained considering an "energy quantum" used by the transporter (which has the same starting term), as could be seen in the right panels of Figure 8 and Figure 7. These results could enforce the biological hypothesis that, instead of ATP, a NH_4^+ concentration gradient (possibly created by the fungus) is used as energy source by the LjAMT2;2 protein.

The simulations which also consider the fungal counterpart are interesting because they provide an initial investigation of this rather poorly characterized side of the symbiosis and confirm that plants can efficiently gain ammonium if NH_4^+ is released from the fungi. This evidence supports the latest biological hypothesis about how fungi supply nitrogen to plants [18,11], and could lead to further models which could suggest which is the needed rate for NH_4^+ transport from fungi to the interfacial apoplast; thus driving biologists toward one (or some) of the nowadays considered hypotheses (active transport of NH_4^+, vesicle formation, etc.).

The explicit representation of pH, while still yielding results comparable with the first simulations, like the pH-dependency of the uptake rate which is clearly represented in Figure 12, offers a scenario where it is possible to analyze more deeply LjAMT2;2 characteristics and the delicate interactions between different transporters.

In the LjAMT2;2 and 1;1 comparison simulations the periarbuscolar pH is mantained at around 4.5 by the previously discussed rules, but by examining H^+ and OH^- one could note that LjAMT2;2 determines an increment of hydrogen which is higher than the one determined by LjAMT1;1, while hydroxide shows oscillations around a mean value for both simulations. This result could suggest that LjAMT2;2 indeed has a role in maintaining the H^+ gradient which is pivotal for other nutrient exchanges that take place in the symbiosis and that its overexpression in the arbusculated cells has a functional meaning.

5   Conclusions and Future Work

This paper reports on the use of SCWC to simulate and understand biological behaviour which is still unclear to biologists, and it also constitues a first attempt to predict biological behaviour and give some directions to biologists for future experiments.

SCWC has been particularly suitable to model the symbiosis (where substances flow through different cells) thanks to its feature of modelling compartments, and their membranes, in a simple and natural way. Our simulations have confirmed some of the latest experimental results about the LjAMT2;2 trans-porter [20] and also support some of the hypotheses about the energy source for the transport and the meaning of LjAMT2;2 overexpression in arbusculated cells. These are the first steps towards a complete simulation of the symbiosis and open some interesting paths that could be followed to better understand the nutrient exchange.

As demonstrated by heterologous complementation experiments in yeast, mycorrhiza-specific plant transporters [22,20] show different uptake efficiencies under varying pH conditions. Looking on this and the results from the simula-tions with different pH conditions in the periarbuscular space new experiments for a determination of the uptake kinetics (K_m-values) under a range of pH seems mandatory. Another conclusion from the model is that, for accurate simulation, exact in vivo concentration measurements have to be carried out even though at the moment this is a difficult task due to technical limitations.

As shown by various studies, many transporters on the plant side show a strong transcriptionally regulation and the majority of them are thought to be localized at the plant-fungus interface [19]. Consequently a quantification of these proteins in the membrane will be a prerequisite for an accurate future model.

It is known that some of these transporters (e.g. PT4 phosphate transporters) obtain energy from proton gradients established by proton pumps (Figure 3) whereas others as the LjAMT2;2 and aquaporines [16] use unknown energy sup-plies or are simply facilitators of diffusion events along gradients. Thus they conserve the membranes electrochemical potential for the before mentioned pro-ton dependent transport processes. To make the story even more complex some of these transporters, which are known to be regulated in the AM symbiosis (e.g. aquaporines), show overlaps in the selectivity of their substrates [25,26,45,33]. Consequently the integration of the involved transporters will be a necessary and challenging task in a future model. It is quite probable that also transporters for other macronutrients (potassium and sulfate) which might be localized in the periarbuscular membrane influence the electrochemical gradients for the known transport processes.

With the ongoing sequencing work [29] on the arbuscular model fungus *Glo-mus intraradices* transporters on the fungal side of the symbiotic compartment are likely to be identified and characterized soon. Data from such future ex-periments could be integrated in the model and help to answer the question whether transporter mediated diffusion or vesicle based excretion events lead to the release of ammonium into the periarbuscular space [11].

Further questions about the plant nutrient uptake and competitive fungal reimport process [11,6] might be answered. Based on the transport properties of orthologous transporters from different fungal and/or plant species, theories could be developed which explain different mycorrhiza responsiveness of host plants; meaning why certain plant-AM fungus combinations have rather a dis-advantageous than a beneficial effect for the plant.

In future research on AM and membrane transport processes in general values from measurements of concentrations or kinetics will be included in the model and will show how the whole system is influenced by these values. Vice versa simulations could be the base for new hypotheses and experiments. As a future extension of the modelling technique, we plan to follow the direction taken in [14,5] for SCLS and to define a type systems to guarantee that an SCWC term satisfies certain biological properties (enforced by a set of typing rules), and also investigate how type systems could enrich the study of quantitative systems.

Acknowledgements. We thank the anonymous CompMod 2009 referees for insightful comments and suggestions on an earlier version this paper.

References

1. Abate, A., Bai, Y., Sznajder, N., Talcott, C., Tiwari, A.: Quantitative and probabilistic modeling in Pathway Logic. In: IEEE 7th International Symposium on Bioinformatics and Bioengineering. IEEE, Los Alamitos (2007)
2. Alberts, B., Johnson, A., Lewis, J., Raff, M., Roberts, K., Walter, P.: Molecular Biology of the Cell. Garland Science (2007)
3. Aldinucci, M., Coppo, M., Damiani, F., Drocco, M., Giovannetti, E., Grassi, E., Troina, A.: SCWC-bio-simulator. Dipartimento di Informatica, Università di Torino (2010), `http://scwc-sim.sourceforge.net/`
4. Alur, R., Belta, C., Ivancic, F.: Hybrid modeling and simulation of biomolecular networks. In: Di Benedetto, M.D., Sangiovanni-Vincentelli, A.L. (eds.) HSCC 2001. LNCS, vol. 2034, pp. 19–32. Springer, Heidelberg (2001)
5. Aman, B., Dezani-Ciancaglini, M., Troina, A.: Type disciplines for analysing biologically relevant properties. Electr. Notes Theor. Comput. Sci. 227, 97–111 (2009)
6. Balestrini, R., Gomez-Ariza, J., Lanfranco, L., Bonfante, P.: Laser microdissection reveals that transcripts for five plant and one fungal phosphate transporter genes are contemporaneously present in arbusculated cells. Mol. Plant Microbe Interact. 20, 1055–1062 (2007)
7. Barbuti, R., Maggiolo-Schettini, A., Milazzo, P., Tiberi, P., Troina, A.: Stochastic calculus of looping sequences for the modelling and simulation of cellular pathways. In: Priami, C. (ed.) Transactions on Computational Systems Biology IX. LNCS (LNBI), vol. 5121, pp. 86–113. Springer, Heidelberg (2008)
8. Barbuti, R., Maggiolo-Schettini, A., Milazzo, P., Troina, A.: Bisimulation congruences in the calculus of looping sequences. In: Barkaoui, K., Cavalcanti, A., Cerone, A. (eds.) ICTAC 2006. LNCS, vol. 4281, pp. 93–107. Springer, Heidelberg (2006)
9. Barbuti, R., Maggiolo-Schettini, A., Milazzo, P., Troina, A.: A calculus of looping sequences for modelling microbiological systems. Fundam. Inform. 72(1-3), 21–35 (2006)
10. Cardelli, L.: Brane calculi. In: Danos, V., Schachter, V. (eds.) CMSB 2004. LNCS (LNBI), vol. 3082, pp. 257–278. Springer, Heidelberg (2005)
11. Chalot, M., Blaudez, D., Brun, A.: Ammonia: a candidate for nitrogen transfer at the mycorrhizal interface. Trends in Plant Science 11, 263–266 (2005)
12. Coppo, M., Damiani, F., Drocco, M., Grassi, E., Troina, A.: Stochastic Calculus of Wrapped Compatnents. In: QAPL 2010. EPTCS, vol. 28, pp. 82–98 (2010)
13. Danos, V., Laneve, C.: Formal molecular biology. Theor. Comput. Sci. 325(1), 69–110 (2004)
14. Dezani-Ciancaglini, M., Giannini, P., Troina, A.: A type system for required/excluded elements in CLS. In: DCM 2009. EPTCS, vol. 9, pp. 38–48 (2009)

15. Gillespie, D.: Exact stochastic simulation of coupled chemical reactions. J. Phys. Chem. 81, 2340–2361 (1977)
16. Gonen, T., Walz, T.: The structure of aquaporins. Q. Rev. Biophys. 39, 361–396 (2006)
17. Gout, E., Bligny, R., Douce, R.: Regulation of intracellular pH values in higher plant cells. Carbon-13 and phosphorus-31 nuclear magnetic resonance studies. J. Biol. Chem. 267, 13903–13909 (1992)
18. Govindarajulu, M., Pfeffer, P.E., Jin, H., Abubaker, J., Douds, D.D., Allen, J.W., Bucking, H., Lammers, P.J., Shachar-Hill, Y.: Nitrogen transfer in the arbuscular mycorrhizal symbiosis. Nature 435, 819–823 (2005)
19. Guether, M., Balestrini, R., Hannah, M.A., Udvardi, M.K., Bonfante, P.: Genome-wide reprogramming of regulatory networks, transport, cell wall and membrane biogenesis during arbuscular mycorrhizal symbiosis in Lotus japonicus. New Phytol. 182, 200–212 (2009)
20. Guether, M., Neuhauser, B., Balestrini, R., Dynowski, M., Ludewig, U., Bonfante, P.: A mycorrhizal-specific ammonium transporter from Lotus japonicus acquires nitrogen released by arbuscular mycorrhizal fungi. Plant Physiology 150, 73–83 (2009)
21. Guttenberger, M.: Arbuscules of vesicular-arbuscular mycorrhizal fungi inhabit an acidic compartment within plant roots. Planta 211, 299–304 (2000)
22. Harrison, M.J., Dewbre, G.R., Liu, J.: A phosphate transporter from *medicago truncatula* involved in the acquisition of phosphate released by arbuscular mycorrhizal fungi. Plant Cell. 14, 2413–2429 (2002)
23. Hause, B., Fester, T.: Molecular and cell biology of arbuscular mycorrhizal symbiosis. Planta 221, 184–196 (2005)
24. Hohnjec, N., Perlick, A.M., Puhler, A., Kuster, H.: The Medicago truncatula sucrose synthase gene MtSucS1 is activated both in the infected region of root nodules and in the cortex of roots colonized by arbuscular mycorrhizal fungi. Mol. Plant Microbe Interact. 16, 903–915 (2003)
25. Holm, L.M., Jahn, T.P., Moller, A.L., Schjoerring, J.K., Ferri, D., Klaerke, D.A., Zeuthen, T.: NH_3 and NH_4^+ permeability in aquaporin-expressing xenopus oocytes. Pflugers Arch. 450, 415–428 (2005)
26. Jahn, T.P., Moller, A.L., Zeuthen, T., Holm, L.M., Klaerke, D.A., Mohsin, B., Kuhlbrandt, W., Schjoerring, J.K.: Aquaporin homologues in plants and mammals transport ammonia. FEBS Lett. 574, 31–36 (2004)
27. Javot, H., Pumplin, N., Harrison, M.J.: Phosphate in the arbuscular mycorrhizal symbiosis: transport properties and regulatory roles. Plant Cell and Environment 30, 310–322 (2007)
28. Knowles, J.R.: Enzyme-catalyzed phosphoryl transfer reactions. Annu. Rev. Biochem. 49, 877–919 (1980)
29. Martin, F., Gianinazzi-Pearson, V., Hijri, M., Lammers, P., Requena, N., Sanders, I.R., Shachar-Hill, Y., Shapiro, H., Tuskan, G.A., Young, J.P.W.: The long hard road to a completed *Glomus intraradices* genome. New Phytologist 180 (2008)
30. Matsuno, H., Doi, A., Nagasaki, M., Miyano, S.: Hybrid Petri net representation of gene regulatory network. In: Prooceedings of Pacific Symposium on Biocomputing, pp. 341–352. World Scientific Press, Singapore (2000)
31. Milazzo, P.: Qualitative and quantitative formal modelling of biological systems. PhD thesis, University of Pisa (2007)
32. Mills, J.C., Roth, K.A., Cagan, R.L., Gordon, J.I.: DNA microarrays and beyond: completing the journey from tissue to cell. Nat. Cell Biol. 3(8), 943 (2001)

33. Niemietz, C.M., Tyerman, S.D.: Channel-mediated permeation of ammonia gas through the peribacteroid membrane of soybean nodules. FEBS Lett. 465, 110–114 (2000)
34. Parniske, M.: Arbuscular mycorrhiza: the mother of plant root endosymbioses. Nat. Rev. Microbiol. 6, 763–775 (2008)
35. Priami, C.: Stochastic pi-calculus. Comput. J. 38(7), 578–589 (1995)
36. Priami, C., Regev, A., Shapiro, E.Y., Silverman, W.: Application of a stochastic name-passing calculus to representation and simulation of molecular processes. Inf. Process. Lett. 80(1), 25–31 (2001)
37. Păun, G.: Membrane computing. An introduction. Springer, Heidelberg (2002)
38. Regev, A., Shapiro, E.: Cells as computation. Nature 419, 343 (2002)
39. Salvemini, F., Marini, A.M., Riccio, A., Patriarca, E.J., Chiurazzi, M.: Functional characterization of an ammonium transporter gene from lotus japonicus. Gene 270, 237–243 (2001)
40. Schaarschmidt, S., Roitsch, T., Hause, B.: Arbuscular mycorrhiza induces gene expression of the apoplastic invertase LIN6 in tomato (lycopersicon esculentum) roots. J. Exp. Bot. 57, 4015–4023 (2006)
41. Schena, M., Heller, R.A., Theriault, T.P., Konrad, K., Lachenmeier, E., Davis, R.W.: Microarrays: biotechnology's discovery platform for functional genomics. Trends Biotechnol. 17(6), 217–218 (1999)
42. Shacharhill, Y., Pfeffer, P.E., Douds, D., Osman, S.F., Doner, L.W., Ratcliffe, R.G.: Partitioning of intermediary carbon metabolism in vesicular-arbuscular mycorrhizal leek. Plant Physiology 108, 7–15 (1995)
43. Singer, S.J., Nicolson, G.L.: The fluid mosaic model of the structure of cell membranes. Science 175, 720–731 (1972)
44. Solaiman, M.D.Z., Saito, M.: Use of sugars by intraradical hyphae of arbuscular mycorrhizal fungi revealed by radiorespirometry. New Phytologist. 136, 533–538 (1997)
45. Tyerman, S.D., Niemietz, C.M., Bramley, H.: Plant aquaporins: multifunctional water and solute channels with expanding roles. Plant Cell Environ. 25, 173–194 (2002)

Genetically Regulated Metabolic Networks: Gale-Nikaido Modules and Differential Inequalities

Ovidiu Radulescu[1,2], Anne Siegel[3,2], Elisabeth Pécou[4], Clément Chatelain[5], and Sandrine Lagarrigue[6]

[1]Université de Montpellier 2, DIMNP - UMR 5235 CNRS/UM1/UM2, Pl. E. Bataillon, Bat 24, CP 107, 34095 Montpellier Cedex 5, France
[2] INRIA Rennes Bretagne Atlantique, Campus de Beaulieu, 35042 Rennes, France
[3] CNRS - Université de Rennes 1, UMR 6074, IRISA, Campus de Beaulieu, 35042 Rennes, France
[4] Université de Nice Sophia Antipolis - CNRS 6621 (Lab. J.-A. Dieudonné), Parc Valrose, 06108 Nice Cedex 02, France
[5] ENS de Cachan Antenne de Bretagne, Campus de Ker Lann, 35170 Bruz, France
[6] Agrocampus - INRA (UMR Génétique animale), 65 rue de St Brieuc, CS 84215, 35042 Rennes, France

Abstract. We propose an approach to study static properties of metabolic networks with genetic regulation. We base our results on differential inequalities which are constraints on the values of the partial derivatives of the reaction rate functions. The approach uses an iterative elimination method for the steady state equations involving algebraic modules that satisfy the Gale-Nikaido global univalence property. The same method allows to find conditions for unique steady state. In the case of metabolic pathways, partial elimination of variables can produce several alternative models, allowing to compare steady state changes of metabolites with and without genetic regulation.

Keywords: systems biology, metabolic control, genetic regulation, qualitative constraints, univalence property.

1 Introduction

The purpose of this paper is to illustrate a new methodology to analyze biological systems which lack numerical information and whose interactions can be structured according to function or timescales. We derive the conditions under which various static properties of these systems are satisfied and look for their biological interpretation. Our reasonings use information on the topology of the reaction network and on the way products regulate fluxes (inhibition or activation).

This work is motivated by the difficulty to build large quantitative models for physiological processes. Although complex biochemical models contain hundreds of reactions and chemical species, these models are relatively small compared to

C. Priami et al. (Eds.): Trans. on Comput. Syst. Biol. XIII, LNBI 6575, pp. 110–130, 2011.
© Springer-Verlag Berlin Heidelberg 2011

models in combustion which contain thousands of reactions (see[1] for instance). The main reason that renders the study of biochemical networks difficult is not (as one may think) the number of variables, but the lack of information on kinetic parameters and, in many cases, the absence of complete knowledge of the mechanisms. It is thus important to develop symbolic tools for the study of biochemical networks, that do not use numerical information. As an example, given a model of genetically regulated metabolism we would like to know if the effect of a gene knock-out or of a change of the nutritional conditions will be an increase or a decrease in the value of a flux or of a concentration of a metabolite. Also, in order to understand the role of the genetic regulations to the stability and performance of metabolism, we would like to compare control coefficients of models with and without genetic regulations. Finally, we would like to find conditions for uniqueness of the steady state. All these motivations can be summarized by a single general question: how to constrain the sign of changes of steady states of perturbed biochemical models, mainly described by their topology and with no kinetic information?

Let us give more insight on methodological aspects. We derive differential models from the topology of the reaction network. Numeric or kinetic information is not required; this is supplied by a set of qualitative inequalities among derivatives of rate functions: roughly, we describe how reactions rates depend on each product of the system - increasing, decreasing, or independent. In order to analyze changes of steady states under genetic perturbation, we extend to a genetic framework the control coefficients introduced in metabolic control analysis [Fel97].

The main methodological problem we face is the computation of signs of control coefficients. To that matter, we introduce a formal method to relate fluxes and metabolites to inputs, based on the study of steady-state equations. Our method employs a powerful condition for uniqueness of steady states - the Gale-Nikaido theorem. We decompose the system into an increasing hierarchy of systems all having steady-states that match with the steady-states of the full system on the variables that are considered. In other words, steady-states of the full system are computed "step-by-step", allowing first the comparisons between the different steps and, second, the identification of those variables which may be implied in non-uniqueness. We use implicit function theorem to compute the hierarchy of control coefficients.

Our method may be considered as an improvement of two different fields. First, extensive studies have been performed on unique steady-state criteria based on the topology of the network [Tho81, CTF06]. Our method includes some of them, and provides new ones. Second, the field of predictions of reaction networks have been largely explored by flux balance analysis (FBA) [LGP06]. However, the results of such methodologies, depend on the judicious choice of the combination of fluxes to optimize. Furthermore, FBA predictions concern fluxes, not metabolites. The reduction methods and our extensions of metabolic

---

[1] http://kinetics.nist.gov/realfuels/

control allow us to extend the range of prediction, since we can discuss the effect
of perturbations also on metabolites.

2   Formalism

2.1   Constraint Based Modeling

We consider chemical kinetic models defined as follows:

$$\frac{d\mathbf{X}}{dt} = \mathbf{\Phi}(\mathbf{X}, \mathbf{p}) \qquad \mathbf{\Phi}(\mathbf{X}, \mathbf{p}) = \sum_i^r \nu_{\mathbf{i}} R_i(\mathbf{X}, \mathbf{p}),$$

where $\mathbf{X}$ denotes the concentration vector of products X_i; $\mathbf{p} \in \mathbb{R}^q$ stands for
a set of parameters of the system; $\mathbf{\Phi} : \mathbb{R}^n \times \Delta \to \mathbb{R}^n$ where Δ is a compact
subset of $\mathbb{R}^q$, is a differentiable vector field; $\nu_{\mathbf{i}}$ is the stoichiometric vector of the
elementary reaction i; $R_i(\mathbf{X}, \mathbf{p})$ is the rate of elementary reaction i.

Our main assumption about the model will be that **The signs of the par-
tial derivatives** $\frac{\partial R_i}{\partial X_j}$ **are constant and known.** This assumption is true for a
very general class of systems, Michaelis-Menten, also power-law approximations
for enzyme-catalyzed reactions, such as Generalized Mass Action and S-systems
[SV87, KKS+06]. Although non-monotonic rates can be obtained by competitive
regulatory mechanisms one could reasonably expect that more complex mechanisms can be decomposed into simple steps for which the constant sign condition
is fulfilled. Alternatively, one may suppose that the constant sign condition is
fulfilled in arcwise connected open domains of the phase space. For our study of
static properties, namely steady state shifts, it will be enough to consider that
both the initial and the final steady states are inside such a domain [RLS+06].

2.2   Steady States, Sequences of Box Equilibration

A steady state is a solution of the system:

$$\mathcal{S} : \mathbf{\Phi}(\mathbf{X}, \mathbf{p}) = 0 \qquad (1)$$

A partial steady state is a solution of $\mathbf{\Phi_1}(\mathbf{X_1}, \mathbf{X_2}, \mathbf{p}) = 0$ where $\mathbf{\Phi} = (\mathbf{\Phi_1}, \mathbf{\Phi_2})$,
$\mathbf{X} = (\mathbf{X_1}, \mathbf{X_2})$ is an arbitrary splitting of the species in the model.

Our aim is to eliminate some or all variables in order to obtain properties
of the system at steady state. We introduce a concept of box equilibration to
perform the substitution method for non-linear systems of equations.

Box of a system of equations. We call box of the system (1) a subset $\mathbf{X}^{(i)}$
of the set of variables $\mathbf{X}$, such that $\mathbf{X} = (\mathbf{X}^{(i)}, \mathbf{X}^{(e)})$ is a partition of the set
of variables. The variables $\mathbf{X}^{(i)}$, $\mathbf{X}^{(e)}$ are called internal and external variables,
respectively. A complete freedom is allowed to decide which variables are internal
and which are external. This choice may of course be guided by biological reasons,
but also by computational reasons. In this framework, internal variables are

those which are going to be removed from the systems. The elimination process described above ensures redistribution of regulatory effects of internal variables over the remaining part of the system. At first sight, the freedom given in the choice of internal variables is misleading since it may have no dynamical reason. However, this freedom shall be eventually considered as a strength since it allows us to focus on the steady states effects of any set variable of the system over other variable, abstracting from timescale and dynamical simulation viewpoints. As illustrated in the paper, this will allow us to understand better the effect of genetic regulation over the metabolic network, which was impossible with usual reduction methods based on dynamics since they prioritize long timescales (here genetic) to short timescales.

To each partition of the variables, let us consider the corresponding partition of the vector field components $\mathbf{\Phi} = (\mathbf{\Phi}^{(i)}, \mathbf{\Phi}^{(e)})$.

We call <u>box equilibration</u> the elimination of internal variables from the equations defined by the internal part of the vector field: $\mathbf{\Phi}^{(i)}(\mathbf{X}^{(i)}, \mathbf{X}^{(e)}, p) = 0$.

Sequence of box equilibration. After a box equilibration the internal variables can be expressed as functions of the external variables. A <u>sequence of box equilibration</u> is the finite iteration of the following operations:

1. Define $\mathbf{X}_1 = \mathbf{X}$ and $\mathbf{\Phi}_1(\mathbf{X}_1, \mathbf{p}) = \mathbf{\Phi}(\mathbf{X}, \mathbf{p})$. We define $\mathcal{D}_1(\mathbf{p}) = \mathbb{R}_+^n$ as the maximal domain which contains solutions for the equation $\mathbf{\Phi}_1(\mathbf{X}_1, \mathbf{p}) = 0$.
2. At k-th iteration, divide the variables and the vector field components into internal and external parts $\mathbf{X}_k = (\mathbf{X}_k^{(i)}, \mathbf{X}_k^{(e)})$, $\mathbf{\Phi}_k = (\mathbf{\Phi}_k^{(i)}, \mathbf{\Phi}_k^{(e)})$.
3. If the external part is not empty then:
   - Let $\mathcal{D}_{k+1}(\mathbf{p})$ be the largest set of points $\mathbf{X}_k^{(e)}$ such that the equation $\mathbf{\Phi}_k^{(i)}(\mathbf{X}_k^{(i)}, \mathbf{X}_k^{(e)}, \mathbf{p}) = 0$ has at least one solution $\mathbf{X}_k^{(i)}$, when $\mathbf{X}_k^{(e)}$ is considered as a fixed parameter, and such that $(\mathbf{X}_k^{(i)}, \mathbf{X}_k^{(e)}) \in \mathcal{D}_k(\mathbf{p})$. If $\mathcal{D}_{k+1}(\mathbf{p})$ is empty then stop: there is no solution. Thus, solving the equation allows expressing the internal variables as functions of the external variables $\mathbf{X}_k^{(i)} = \mathcal{M}_\mathbf{k}(\mathbf{X}_k^{(e)}, \mathbf{p})$. Notice that the solution might not be unique, that is $\mathcal{M}_\mathbf{k}$ is not necessarily univalent. We restrict our discussion to the case when the number of solutions is finite and bounded, such as for polynomial systems.
   - define $\mathbf{X}_{\mathbf{k}+1} = \mathbf{X}_\mathbf{k}^{(\mathbf{e})}$, and $\mathbf{\Phi}_{\mathbf{k}+1} = \mathbf{\Phi}_\mathbf{k}^{(\mathbf{e})}(\mathcal{M}_\mathbf{k}(\mathbf{X}_n^{(\mathbf{e})}, \mathbf{p}), \mathbf{X}_n^{(\mathbf{e})}, \mathbf{p})$.
4. If the external part is empty then solve $\mathbf{\Phi}_k^{(i)}(\mathbf{X}_k^{(i)}, \mathbf{p}) = 0$ and stop. Conventionally, in this case $\mathcal{D}_{k+1}(\mathbf{p})$ is considered non-empty iff the equation has a solution.
5. go to step 2.

A sequence of box equilibration is <u>complete</u> if all components are equilibrated i.e.

$$\mathbf{X} = \mathbf{X}_1^{(i)} \oplus \mathbf{X}_2^{(i)} \oplus \ldots \oplus \mathbf{X}_{N_b}^{(i)}.$$

After a complete sequence of box equilibration one should be able to express steady state species concentrations as functions of the external parameters: $\mathbf{X} = \mathcal{M}(\mathbf{p})$, where $\mathcal{M}$ results from a composition of the functions $\{\mathcal{M}_\mathbf{k}\}_{k=1, N_b}$:

$$\mathbf{X_k^{(i)}(p)} = \mathcal{M_k}(\mathbf{X_{k+1}^{(i)}(p)}, \ldots, \mathbf{X_{N_b}^{(i)}(p)}).$$

A well known example of (incomplete) sequence of box equilibration is a slow/fast reduction of a system: fast variables are reduced in order to obtain equivalent steady states for slow variables.

The sequence of equilibration is justified when the existence and uniqueness of solutions to the full system $\boldsymbol{\Phi} = 0$ are not straightforward. In the sequel, we will introduce conditions for existence and uniqueness. For computational (or biological significance) reasons, checking these conditions will be performed on subsets of variables and not on the full system itself. Hence, the concept of sequence of equilibration provides a flexibility in checking uniqueness conditions: we first equilibrate a set of internal variables satisfying uniqueness, perform reduction, then exhibit a new set of variables to be tested for uniqueness... At the end of the process, the set of variables that may be responsible for non-uniqueness of the steady state are eventually identified with this process. Additionally, performing reduction "step-by-step" allows us to derive biological interpretations of the conditions that would not be possible if the reduction would be performed in one step.

Existence and uniqueness of solutions. Box equilibration solves systems of equations by substitution. The existence and uniqueness of solutions relatively to box equilibration are straightforward.

Proposition 2.1. $\quad -$ *A solution of the system* (1) *exists for a value of the parameter p if there is a complete sequence of box equilibration with non-empty domains $\mathcal{D}_{k+1}(\mathbf{p})$.*
  - *The function $\mathcal{M}$ is univalent (to one $\mathbf{p}$ corresponds a single value of $\mathcal{M}$) if all the domains $\mathcal{D}_{k+1}(\mathbf{p})$ are non-empty and each one of the function $\mathcal{M_k}$ is univalent on its maximal domain $\mathcal{D}_{k+1}(\mathbf{p})$ for a complete sequence of box equilibration.*

This property is useful to prove the existence and uniqueness of solutions of systems of non-linear equations. It is enough to choose a complete sequence of box equilibration and to show that at each step the functions $\mathcal{M_k}$ are univalent on non-empty domains $\mathcal{D}_{k+1}(\mathbf{p})$.

It is difficult to give a "only if" version of the property. Indeed, even if we find a box such that the equations $\boldsymbol{\Phi}_k^{(i)}(\mathbf{X}_k^{(i)}, \mathbf{X}_k^{(e)}, \mathbf{p}) = 0$ have multiple solutions in $\mathbf{X}_k^{(i)}$ it is not excluded that some of these solutions are incompatible with the rest of the equations: after all the box equilibration we may still have an unique solution.

2.3   Checking Box Equilibration: Existence of Partial Steady States

Let us state a sufficient condition for existence of steady states.

Theorem 2.1. *Let $\boldsymbol{\Phi}(\mathbf{X}) = \mathbf{G}(\mathbf{X}) - \boldsymbol{\Lambda}(\mathbf{X})$ be a smooth vector field on $\mathbb{R}_+^n$ ($\mathbb{R}_+^n$ represents all the vectors of $\mathbb{R}^n$ having non-negative coordinates) such that :*

1. $\mathbf{G}$ *is bounded,*
2. *For all* $\mathbf{X} = (X_1, \ldots, X_n)$ *such that* $X_i = 0$ *and* $X_j \neq 0$ *for all* $j \neq i$, $\mathbf{G}$ *satisfies* $G_i(\mathbf{X}) > 0$,
3. $\mathbf{\Lambda} = (\Lambda_1(X_1), \ldots, \Lambda_n(X_n)) : \mathbb{R}_+^n \to \mathbb{R}_+^n$, *and* Λ_i *are differentiable and satisfy* $\Lambda_i(0) = 0$ *and* $\lim_{\|\mathbf{X}\| \to +\infty} \Lambda_i(\mathbf{X}) = +\infty$, *for all* $1 \leq i \leq n$.

Then the equation $\mathbf{\Phi}(\mathbf{X}) = 0$ *has at least one solution in* $\mathbb{R}_+^n$.

The proof of the Theorem 2.1 is based on the following standard mathematical lemma which is a consequence of the Poincaré-Hopf formula (see Additional material in the Appendix).

Lemma 2.1. *Let* D *be a smooth ball in* $\mathbb{R}^n$ *and let* S *be the boundary of* D. *Let* Φ *be a differentiable vector field defined on a neighborhood of* D. *If* Φ *points inward* D *at any point of* S *then* $\mathbf{\Phi}$ *admits a zero in the interior of* D.

Biological interpretation. This theorem is very general but it can be considered quite easily in a biological setting: $\mathbf{G}$ may be considered as the result of production and consumption of products by biochemical reactions with a saturation effect, whereas $\mathbf{\Lambda}$ represents the (unbounded) product degradation. More precisely, the hypotheses of Theorem 2.1 are fulfilled by rather general networks of biochemical reactions, by assuming the following rules:

- For each variable X_i, degradation terms $\mathrm{Dg}(X_i)$ are increasing function of X_i with no saturation effect.
- In the absence of substrates all fluxes vanish.
- All fluxes except degradation saturate at high concentrations of metabolites, implying that production terms $\mathbf{\Phi}$ are bounded.
- There exists a recovery effect on each metabolic variable. By recovery effect we mean that if a variable is zero, then at least one reaction that produces the variable is active.

The following consequence of Theorem 2.1 is very useful to exhibit complete sequences of box equilibration.

Corollary 2.1. *Let* $\mathbf{X} = (\mathbf{X_1}, \mathbf{X_2})$, $\Phi = (\Phi_1, \Phi_2)$ *be any partition of the variables. We suppose that* $\Phi_1(\mathbf{X_1}, \mathbf{X_2})$ *satisfies the hypotheses of Theorem 2.1, as a function of* $\mathbf{X_1}$. *Then, given* $\mathbf{X_2}$, *the system of equations* $\Phi_1(\mathbf{X_1}, \mathbf{X_2}) = 0$, *where* $\mathbf{X_2}$ *is considered as a constant parameter vector, admits a solution in* $\mathbf{X_1}$ *with non negative entries.*

2.4   Checking Box Equilibration: Uniqueness of Partial Steady State

In order to identify boxes of equilibration, we introduce an algebraic sufficient condition on signs of derivatives for uniqueness. We use the following result which is a direct consequence of Gale-Nikaido-Inada theorem [Par83]. This theorem can be seen as a generalization to higher dimensions of the monotonicity of functions on $\mathbb{R}$. Let us recall that a principal minor of a matrix $M = (m_{i,j})_{i,j \in \{1,\ldots,n\}}$ is defined as $\Delta_I = \det M_I$, where $I \subset \{1, \ldots, n\}$ and $M_I = (m_{i,j})_{i,j \in I}$.

Theorem 2.2 (Gale-Nikaido). *If* $X \to \Phi(X)$ *is a differentiable mapping from* $\mathbb{R}_+^n$ *to* $\mathbb{R}^n$, *of Jacobian* J, *such that all the principal minors of* $-J$ *are positive, then this mapping is globally univalent. In particular the system* $\Phi = 0$ *has a unique solution if a solution exists.*

Recovering Thomas condition for uniqueness of steady states. Let us detail why this result can be seen as a generalization of the well known Thomas condition for uniqueness of the steady state. The interaction graph is the signed, oriented graph on the variables derived from the Jacobian J of the model: j is connected to i ($j \to i$) iff $J_{ij} \neq 0$. The sign of the connecting arc is the sign of J_{ij}. Many known topological conditions for uniqueness of steady state actually follow from the cycle decomposition of the determinant of $-J$, that is, $\Delta(-J) = \sum_{L \in \mathcal{L}} (-1)^{|L|} lp(L)$, where $\mathcal{L}$ is the set of all cycle partitions and $|L|$ is the number of cycles in the partition L, $lp(L)$ is the product of elements J_{ij} for all the arcs in L. As a particular case, assume that there are no positive cycles in the interaction graph (Thomas sufficient condition for uniqueness of steady state [Tho81]). Consider a principal minor J and the corresponding subgraph $\mathcal{G}$. Let L be a partition of $\mathcal{G}$ into $|L|$ disjoint cycles $l_1 \ldots l_{|L|}$. Each cycle l_k is also a cycle of the full interaction graph, implying that the product of signs in each cycle l_k is negative. Therefore, the sign $lp(L)$ of arcs in the partition L equals $(-1)^{|L|}$. Summing up these relation yields $\Delta(-J) = \sum_{L \in \mathcal{L}} (-1)^{|L|} lp(L) = \sum_{L \in \mathcal{L}} (-1)^{2|L|} = |\mathcal{L}| > 0$. From Theorem 2.2 we get the uniqueness of the steady state.

Limitations of Thomas conditions. Nevertheless, Thomas condition is too restrictive for most of the applications, especially to metabolism. Indeed, a reversible reaction can be represented as a set of two reactions of opposite stoichiometry: $X_i \to X_j$ and $X_j \to X_i$ of rates $R(X_i, X_j)$, $R'(X_j, X_i)$, respectively. Since X_i, X_j are substrates of R, R', we have $\frac{\partial R}{\partial X_i} > 0$ and $\frac{\partial R'}{\partial X_j} > 0$ and it follows that $i \to j \to i$ is a positive loop in the interaction graph; Thomas condition does not apply.

Let us go further in this example. We notice that contribution of reactions R, R' to the decomposition of $\Delta(-J)$ is zero, because the contribution to any cycle partition containing $i \to j \to i$ is exactly canceled by the contribution to the cycle partition containing $i \to i$ and $j \to j$. Actually, checking formally the Gale-Nikaido condition allows us to avoid the limitations of the Thomas condition.

A reversible reaction corresponds to a non-essential loop in the interaction graph, i.e. a loop not contributing to multi-stationarity. In this paper we check uniqueness of steady state by direct application of the Gale-Nikaido condition. Non-essential loops have vanishing contribution to principal minors of the Jacobian and are thus automatically eliminated by this procedure.

3   Constraint Based Model for Fatty Acids Metabolism

We apply our formalism to a minimal mixed metabolic and genetic model of regulated fatty acids metabolism in liver. To set ideas, all the variables of the model pertain to an "abstract" hepatocyte, capable of the two different functioning modes. We thus voluntarily reduced the set of elements in the model. These belongs to three different classes of molecules. Their corresponding symbols are given in Table 1.

- **Parameter:** The system is driven by the glucose concentration, representing food. Different nutritional states such as normal feeding or fasting are modeled by different values of this parameter.
- **Metabolic variables:** Acetyl-CoA is the first brick for building fatty acids; Saturated and monounsaturated fatty acids (denoted by S/MU-FA) are produced either by the organisms from Acetyl-CoA or brought by the diet; Exogenous polyunsaturated fatty acids (PUFA) are entering the metabolism only as part of the diet.
- **Energetic variable:** a variable ATP expresses the energy that the cell has at its disposal.
- **Genetic variables:** we introduce abstract enzymes for each set of enzymes that are involved in a metabolic pathway and main transcription factors known to regulate these enzymes, namely, the active form of the nuclear receptor PPAR and the active form of the nuclear receptor LXR, representing in a very simplified way the regulation path LXRα-SREBP-1.

In different species such as poultry, rodents and humans, hepatocyte (liver) cells have the specificity to ensure both lipogenesis and β-oxidation. We thus abstracted the main fluxes and regulations implied in this biological process.

- **Metabolic fluxes.** Glycolysis produces Acetyl-CoA from glucose. Krebs cycle produces energy for cellular needs from Acetyl-CoA. Ketone bodies exit allows the cell to transfer the energy stored in Acetyl-CoA to the outside, allowing survival during fasting. Lipogenesis transforms Acetyl-CoA into S/MU-FA via citrate. Then an outtake flux allows S/MU-FA to exit liver and go to storing tissues (adipocytes); this flux is reversible since the intake flux is fed partially from diet, partially from lipolysed adipocytes. Additionally, the intake/outtake flux of PUFA allows PUFA to enter or exit the cell, including a synthetic pathway consisting of desaturation and elongation of essential fatty acids. When fatty acids enter the cell, a $\beta-$oxidation burns all fatty acids in order to produce energy and to recover Acetyl-CoA. Finally, ATP consumption expresses the energy (ATP) the cell consumes for living. Degradation of metabolites can not be neglected on the genetic timescale.
- **Genetic regulations.** Fluxes are regulated by their sets of enzymes, which are themselves regulated by transcription factors PPARα and LXRα. More precisely, LXR and SREBP-1 triggers S/MU-FA synthesis enzymes production and PPAR triggers the production of S/MU-FA oxidation enzymes, PUFA oxidation enzymes and ketone exit enzymes.
- **Activity regulations.** It has been established that fatty acids can upregulate or down-regulate the expression of different genes controlling their metabolism. The regulatory effect is mainly due to PUFA (see details in

Table 1. (Left) The <u>full model for fatty acid metabolism</u>. Dashed arrows stand for genetic actions from the origin on to target. Plain arrows stand for metabolic fluxes. Dash-dot arrows stand for energetic regulations implying T. In this model, notice that a metabolite F_2 (that is, polyunsaturated fatty acids PUFA) regulate the genetic regulators L (LXRα-SREBP-1 pathway) and PP (PPAR-α).
(Right) Differential equations for the full model. The flux of each metabolic variable is obtained as a mass balance of primitive fluxes.

Type	Name	Concentration symbol	$\left\|\dfrac{d\,product}{dt}\right\|$
Metabolic parameter	Glucose	G	
Metabolic variable	Acetyl Co-A	A	Φ_A
	Saturated and monounsaturated fatty acids (S/MU-FA)	F_1	Φ_{F_1}
	Poly-unsaturated fatty acids (PUFA)	F_2	Φ_{F_2}
Energetic variable	Energy ATP	T	Φ_T
Genetic variable	Active form of PPAR	PP	Ψ_1
	Active form of the regulation path LXR-SREBP	L	Ψ_2
	Enzymes of S/MU-FA synthesis	E_1	Ψ_3
	Enzymes of S/MU-FA oxidation	E_2	Ψ_4
	Enzymes of PUFA oxidation	E_3	Ψ_5
	Enzymes of Ketone body exit	E_4	Ψ_6

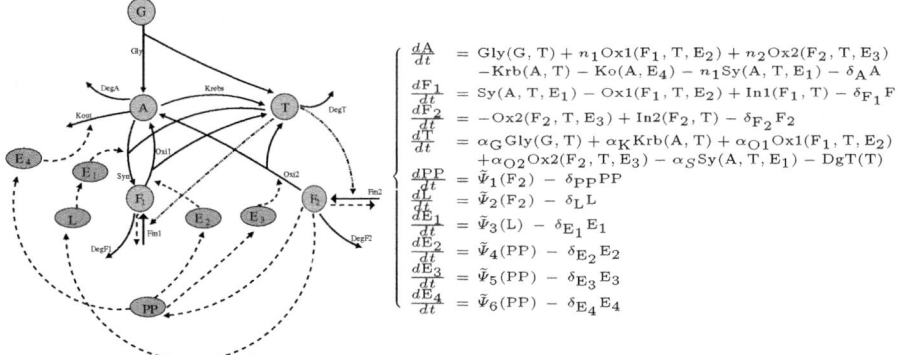

$$\begin{cases}
\dfrac{dA}{dt} = \mathrm{Gly}(G,T) + n_1\mathrm{Ox1}(F_1,T,E_2) + n_2\mathrm{Ox2}(F_2,T,E_3) \\
\qquad\quad - \mathrm{Krb}(A,T) - \mathrm{Ko}(A,E_4) - n_1\mathrm{Sy}(A,T,E_1) - \delta_A A \\
\dfrac{dF_1}{dt} = \mathrm{Sy}(A,T,E_1) - \mathrm{Ox1}(F_1,T,E_2) + \mathrm{In1}(F_1,T) - \delta_{F_1} F_1 \\
\dfrac{dF_2}{dt} = -\mathrm{Ox2}(F_2,T,E_3) + \mathrm{In2}(F_2,T) - \delta_{F_2} F_2 \\
\dfrac{dT}{dt} = \alpha_G\mathrm{Gly}(G,T) + \alpha_K\mathrm{Krb}(A,T) + \alpha_{O1}\mathrm{Ox1}(F_1,T,E_2) \\
\qquad\quad + \alpha_{O2}\mathrm{Ox2}(F_2,T,E_3) - \alpha_S\mathrm{Sy}(A,T,E_1) - \mathrm{DgT}(T) \\
\dfrac{dPP}{dt} = \tilde{\Psi}_1(F_2) - \delta_{PP}PP \\
\dfrac{dL}{dt} = \tilde{\Psi}_2(F_2) - \delta_L L \\
\dfrac{dE_1}{dt} = \tilde{\Psi}_3(L) - \delta_{E_1} E_1 \\
\dfrac{dE_2}{dt} = \tilde{\Psi}_4(PP) - \delta_{E_2} E_2 \\
\dfrac{dE_3}{dt} = \tilde{\Psi}_5(PP) - \delta_{E_3} E_3 \\
\dfrac{dE_4}{dt} = \tilde{\Psi}_6(PP) - \delta_{E_4} E_4
\end{cases}$$

[CF03, PLM03, DF04, Jum04]). Although the precise mechanisms have not been proved yet, some well established facts are used for modeling: PUFA increases PPAR activity and inhibits LXR activity.

- **Energetic regulations on fat intake.** Fat intake is needed to produce energy by oxidation. A drop in energy (ATP) stimulates fat intake.

Our full model for fatty acid metabolism and its regulations is depicted in Table 1. It was built from these interactions by using the following rules.

- The production Φ_A, Φ_{F_1}, Φ_{F_2}, Φ_T of each metabolic variable is obtained as the sum of primitive fluxes that produce or consume the metabolite.
- Primitive fluxes are treated as single reactions with simple stoichiometry. Thus, the fluxes Gly, Krb, Ox1, Ox2, Sy are considered to have the stoichiometry's $G \to A + \alpha_G T$, $A \to \alpha_K T$, $F_1 \to n_1 A + \alpha_{O1} T$, $F_2 \to n_1 A + \alpha_{O2} T$, $n_1 A + \alpha_S T \to F_1$ respectively.
- Degradation reactions of metabolites are supposed to be linear: $\mathrm{DgV}(V) = \delta_V V$ where V denotes any variable A, F_1, F_2.

Table 2. Constrained signs on the full model for regulated fatty acid metabolism in liver.

$\frac{\partial\,\text{flux}}{\partial\,\text{var.}}$	Gly	Krb	Ko	Sy	Ox1	Ox2	In1	In2	DgT	$\tilde{\Psi}_1$	$\tilde{\Psi}_2$	$\tilde{\Psi}_3$	$\tilde{\Psi}_4$	$\tilde{\Psi}_5$	$\tilde{\Psi}_6$
A	0	+	+	+	0	0	0	0	0	0	0	0	0	0	0
F_1	0	0	0	0	+	0	−	0	0	0	0	0	0	0	0
F_2	0	0	0	0	0	+	0	−	0	+	−	0	0	0	0
T	−	−	0	+	−	−	−	+	0	0	0	0	0	0	0
PP	0	0	0	0	0	0	0	0	0	0	0	0	+	+	+
L	0	0	0	0	0	0	0	0	0	0	0	+	0	0	0
E_1	0	0	0	+	0	0	0	0	0	0	0	0	0	0	0
E_2	0	0	0	0	+	0	0	0	0	0	0	0	0	0	0
E_3	0	0	0	0	0	+	0	0	0	0	0	0	0	0	0
E_4	0	0	+	0	0	0	0	0	0	0	0	0	0	0	0
G	+	0	0	0	0	0	0	0	0	0	0	0	0	0	0

– The functions Ψ_i expressing variations of the genetic variables (PP, L, E_1, E_2, E_3, E_4) were not detailed because mechanisms are still unknown. Instead, each function Ψ_i has been decomposed into a non-negative production term $\widetilde{\Psi}_i$ and a linear degradation term.

The information already given about regulations allows to partially fill the table of partial derivatives of the fluxes. To identify the remaining signs, we use the following assumptions: a) Substrate effect, an increase of substrate increases the associated flux; b) Transport effects, intake/outtake fluxes In1 and In2 are conventionally directed to the inside, they decrease when the internal concentrations of fatty acids increase; c) Product negative feed-back, fluxes producing ATP are negatively controlled by ATP. The resulting table of derivation is given in Table 2.

4 Results

We apply the results detailed in the first section to derive several information about fatty acid metabolism.

4.1 Two Models of Response of the Metabolism

Genetically non-regulated model. A genetically regulated system is multi-scale. During fast response genetic variables can be considered to be constant and equal to their initial values. We call genetically non-regulated model the reduced model obtained from the full model by considering $E_i(t) = E_i(0), i = 1,\ldots,4$, $\text{PP}(t) = \text{PP}(0)$, $\text{L}(t) = \text{L}(0)$. Fast response is obtained at partial steady state of the remaining four variables $\{A, F_1, F_2, T\}$. By construction, the genetically non-regulated model can be used to describe the rapid response of the metabolic variables on timescales smaller than the relaxation time of the genetic variables. In particular, the steady state of this model are the quasi-stationary states of the full model.

The genetically regulated model. Slow adiabatic response involves all variables, including genetic ones. Nonetheless, the static response of the system can be obtained by arbitrarily choosing the order of partial equilibration. In order to study the impact of genetic regulation on energetic homeostasis, we consider a second reduced model involving the same four variables. In the full model the box formed by the species $\{E_1, E_2, E_3, E_4, PP, L\}$ is acyclic and satisfies the conditions of Corollary 2.1, thus partial steady state exists and is unique. We call genetically regulated model the reduction of the full model to the set $\{A, F_1, F_2, T\}$ (obtained by elimination of the genetic variables). The chain rule formula allows to calculate the signs of flux derivatives obtained after reduction. The genetically regulated model and the full model do not simulate the same dynamics, but their steady states are identical over the set of variables $\{A, F_1, F_2, T\}$.

The genetically regulated model and the genetically non-regulated model have the same structure in terms of variables, parameters and fluxes. However, performing reductions affects the dependencies of genetically regulated fluxes Ko, Sy, Ox1, Ox2, In1, In2 on other variables. The value of these fluxes after elimination is denoted by subscripts gr (standing for genetically regulated) or gnr (for non genetically regulated), for instance Ko_{gr} or Ko_{gnr}. The models are detailed in Table 3 together with the corresponding sign table. As shown in the table, the only difference is in the regulation: some fluxes do not depend on F_2 in the genetically non-regulated model, since the regulations of F_2 are different in the genetically non-regulated and genetically regulated models, leading to two different functions F_2^{gnr} and F_2^{gr}.

Notice that the two models models do not simulate the same dynamics and that their steady states match, on the metabolic variables, with the value of quasi-stationary and steady states of the full model, respectively. Comparing the steady states of these two models will allow a characterization of the effect of genetic regulations on the full model.

Control coefficients and elasticities. We call control coefficients the derivatives of fluxes with respect to F_2 and T. We also call elasticity the derivative of the logarithm of the rate of metabolic variables with respect to the logarithm of the substrate concentration: they quantify how rates and fluxes of a metabolite depend on this metabolite [CB95]. Corresponding symbols are given in Table 3. These quantities are defined such that they are all positive.

4.2   Condition for Unique Steady State

We can now turn to the application of our theoretical results about existence and uniqueness of equilibria. First, let us notice that hypotheses of Theorem 2.1 apply to the full model and to the reduced models. All these models admit at least a steady state.

Bistability of genetically regulated metabolism is used by some organisms to adapt to a change in food (see the operon lactose in E.coli). There are two functioning antagonist modes of the fatty acid metabolism in liver: lipogenesis that produce reserves, fatty acid oxidation that burns reserves and produces energy. The choice of the functioning mode depends on nutrition conditions: a lack of food (i.e. a sustained low level of glucose) stimulates lipolysis and oxidation; normal feed (normal glucose level) induces lipogenesis. This motivates the first biological

Table 3. The genetically non-regulated and genetically regulated models given as differential equations and graph. The differential equations simulate the correct dynamics only for the genetically non-regulated model, and only on the rapid part of the trajectory before reaching quasi-stationarity. For genetically regulated model they only provide the same steady state as the full model, a condition which is sufficient for the study of static properties. The table of signs contains symbols corresponding to control coefficients and elasticities. Notations *gnr* and gr denote the value of fluxes for the different models (non-genetically or genetically regulated). All the symbols stand for positive values. These quantities inform on the strength of fluxes variations one with respect to the other and how rates and fluxes of a metabolite depend on this metabolite. They will be used in the sequel to express conditions on the system to satisfying specific behaviors.

$$
\begin{cases}
\frac{dA}{dt} = -\delta_A A + \text{Gly}(G, T) + n_1 \text{Ox1}_{gr,gnr}(F_1, F_2, T) + n_2 \text{Ox2}_{gr,gnr}(F_2, T) - \text{Krb}(A, T) - \\
\qquad \text{Ko}_{gr,gnr}(A, F_2) - n_1 \text{Sy}_{gr,gnr}(A, F_2, T) \\
\frac{dF_1}{dt} = \text{Sy}_{gr,gnr}(A, F_2, T) - \text{Ox1}_{gr,gnr}(F_1, F_2, T) + \text{In1}(F_1, T) - \delta_{F_1} F_1 \\
\frac{dF_2}{dt} = -\text{Ox2}_{gr,gnr}(F_2, T) + \text{In2}(F_2, T) - \delta_{F_2} F_2 \\
\frac{dT}{dt} = \alpha_K \text{Krb}(A, T) + \alpha_{O1} \text{Ox1}_{gr,gnr}(F_1, F_2, T) + \alpha_{O2} \text{Ox2}_{gr,gnr}(F_2, T) - \alpha_S \text{Sy}_{gr,gnr}(A, F_2, T) \\
\qquad +\alpha_G \text{Gly}(G, T) - \text{DgT}(T)
\end{cases}
$$

$\frac{\partial \Phi}{\partial X}$		Gly	Krb	Ko	Sy	Ox1	Ox2	In1	In2	DgT
	A	0	χ_A^{Krb}	+	χ_A^{Sy}	0	0	0	0	0
	F_1	0	0	0	0	$\chi_{F_1}^{\text{Ox1}}$	0	−	0	0
$F_2 \begin{matrix}gnr\\gr\end{matrix}$		0	0	$\begin{matrix}0\\R_{F_2}^{\text{Ko}}\end{matrix}$	$\begin{matrix}0\\-R_{F_2}^{\text{Sy}}\end{matrix}$	$\begin{matrix}0\\R_{F_2}^{\text{Ox1}}\end{matrix}$	$R_{F_2}^{\text{Ox2}}$	0	−	0
	T	$-R_T^{\text{Gly}}$	$-R_T^{\text{Krb}}$	0	R_T^{Sy}	$-R_T^{\text{Ox1}}$	$-R_T^{\text{Ox2}}$	$-R_T^{\text{In1}}$	$-R_T^{\text{In2}}$	+
	G	+	0	0	0	0	0	0	0	0

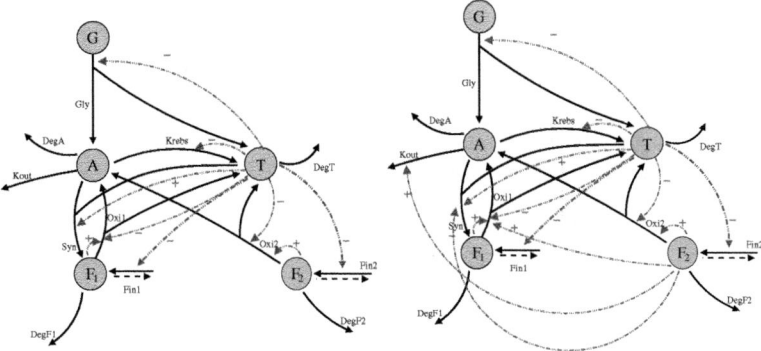

a) genetically non-regulated model.   b) genetically regulated model.

question we wish to answer: in higher organisms, does the whole of regulations produce bistability or a unique steady state of fatty acid metabolism, or is the change from lipogenesis to lipolysis merely steady state shift?

In order to answer this question, we will use the Gale-Nikaido Theorem 2.2 to exhibit a complete sequence of univalent equilibration: the first box is $\{A, F_1, F_2\}$ and the second box $\{T\}$. This will allow us to exhibit an algebraic condition for unique steady state. The biological relevance of this condition will be discussed at the end of the section.

To this matter, we introduce new elasticities, all positive:
$$\chi_{F_1}^{tot} = -\frac{\partial \Phi_{F_1}}{\partial F_1}, \qquad \chi_{F_2}^{tot} = -\frac{\partial \Phi_{F_2}}{\partial F_2}, \qquad \chi_A^{tot} = -\frac{\partial \Phi_A}{\partial A}.$$

We can also show that the following ratios are all positive and strictly less than 1:
$$\rho_{F_1}^{Ox1} = \frac{\chi_{F_1}^{Ox1}}{\chi_{F_1}^{tot}}, \quad \rho_A^{Sy} = \frac{n_1 \chi_A^{Sy}}{\chi_A^{tot}}, \quad \rho_{F_2}^{Ox2} = \frac{R_{F_2}^{Ox2}}{\chi_{F_2}^{tot}}$$

Let us define the following combinations of control coefficients and elasticities.

$$
\begin{aligned}
A &= X\rho_{F_1}^{Ox1}, \qquad X = n_1(\alpha_{O1} - \alpha_S \rho_A^{Sy} + n_1 \alpha_K \rho_A^{Krebs}), \\
B &= B_1 R_{F_2}^{Sy}/\chi_{F_2}^{tot} + B_2 R_{F_2}^{Ko}/\chi_{F_2}^{tot} + B_3 R_{F_2}^{O1}/\chi_{F_2}^{tot} + B_4 \rho_{F_2}^{Ox2}, \\
B_1 &= X - n_1(\alpha_{O1} - \alpha_S)(1 - \rho_A^{Sy}\rho_{F_1}^{Ox1}), \qquad B_2 = \alpha_{O1}(1 - \rho_A^{Sy}\rho_{F_1}^{Ox1}) - X/n_1, \\
B_3 &= X(1 - \rho_{F_1}^{Ox1}), \qquad B_4 = n_1 \alpha_{O2}(1 - \rho_A^{Sy}\rho_{F_1}^{Ox1}) + n_2/n_1 X - n_2 \alpha_{O1}(1 - (\rho_A^{Sy})^2 \rho_{F_1}^{Ox1}), \\
C &= [X/n_1 + (n_1 \alpha_G - \alpha_{O1})(1 - \rho_A^{Sy}\rho_{F_1}^{Ox1})]R_T^{Gly} + [\alpha_{O1} + n_1 \alpha_K + X R_T^{Oxi1} \qquad (2) \\
& \quad [n_2/n_1 X + n_2 \alpha_S \rho_A^{Sy}(1 - \rho_{F_1}^{Ox1}) + (n_1 \alpha_{O2} - n_2 \alpha_{O1})(1 - \rho_A^{Sy}\rho_{F_1}^{Ox1})]R_T^{Oxi2} + \\
& \quad -(\alpha_S + X/n_1)\rho_{F_1}^{Ox1} + \rho_A^{Sy}]R_T^{Krebs} + [n_1(\alpha_S - \alpha_{O1}) + X](1 - \rho_{F_1}^{Ox1})R_T^{Syn}, \\
D &= [X/n_1 - \alpha_{O1}(1 - \rho_A^{Sy})]R_T^{Krebs} + n_2 \alpha_S \rho_A^{Sy}(1 - \rho_{F_1}^{Ox1})R_T^{Oxi2} \\
& \quad + n_1(\alpha_{O1} - \alpha_S)(1 - \rho_A^{Sy})\rho_{F_1}^{Ox1} R_T^{S}.
\end{aligned}
$$

As detailed in the proof of Theorem 4.1 above, theses combinations result from box equilibration – Section 2.2 – together with the Gale-Nikaido uniqueness condition – Theorem 2.2 – and the implicit function theorem. All together, we obtain combinations of coefficients that compose the steady-state uniqueness conditions for the models we are considering. The biological interpretation of these coefficients is discussed at the end of the present section.

Theorem 4.1. *Assume that the following strong lipolytic response condition (3) and fatty acid proportion condition (4) are fulfilled at fixed genetic variables and at genetic partial steady state, for every $G \in [0, G_{max}]$. Suppose additionally that that the stoichiometry condition (5) is satisfied:*

$$A(R_T^{In1} - R_T^{Ox1}) + C > D, \qquad (3)$$
$$|B(R_T^{In2} - R_T^{Ox2})| <<< A|R_T^{In1} - R_T^{Ox1}| \qquad (4)$$
$$\alpha_S < \alpha_{O1} < n_1 \alpha_G, \qquad n_2 \alpha_{O1} < n_1 \alpha_{O2}. \qquad (5)$$

Then the model of fatty acid metabolism has a unique steady state, with or without genetic regulation. The quantities A, C, D are positive.

Proof. Let us first prove that for all $(G, T) \in [0, G_{max}] \times \mathbb{R}_+$, the box $\{A, F_1, F_2\}$ can be eliminated from equilibria equations in the genetically non-regulated model and the genetically regulated model. Corollary 2.1 implies that the system of equations $\Phi_A(G, A, F_1, F_2, T) = \Phi_{F_1}(A, F_1, F_2, T) = \Phi_{F_2}(F_2, T) = 0$ has a solution for every fixed (G, T). To prove the uniqueness of the solution to the system, we apply Theorem 2.2 to the mapping $(A, F_1, F_2) \rightarrow (\Phi_A, \Phi_{F_1}, \Phi_{F_2})$; let $J^{(1)}$ is the Jacobian of this mapping:

$$J^{(1)} = \begin{pmatrix} -\chi_A^{tot} & n_1\chi_{F_1}^{Ox1} & n_2 R_{F_2}^{Ox2} + n_1 R_{F_2}^{Ox1} + n_1 R_{F_2}^{Sy} - R_{F_2}^{Ko} \\ \chi_A^{Sy} & -\chi_{F_1}^{tot} & -R_{F_2}^{Ox1} - R_{F_2}^{Sy} \\ 0 & 0 & -\chi_{F_2}^{tot} \end{pmatrix}.$$

We ensure that all the principal minors of $-J^{(1)}$ are all positive: $\chi_A^{tot} > 0$, $\chi_A^{tot}\chi_{F_1}^{tot} - n_1\chi_A^{Sy}\chi_{F_1}^{Ox1} = \chi_A^{tot}\chi_{F_1}^{tot}(1 - \rho_A^{Sy}\rho_{F_1}^{Ox1}) > 0$, $\chi_{F_2}^{tot}\chi_A^{tot}\chi_{F_1}^{tot}(1 - \rho_A^{Sy}\rho_{F_1}^{Ox1}) > 0$, as a consequence of the sign table. This is valid both at fixed genetic variables and at partial steady state of genetic variables.

Alternatively this can be seen from the topology of the reaction network. The only cycle of the box $\{A, F_1, F_2\}$ comes from the pair of opposed fluxes Sy, Ox1; or, these are hanging equations.

Since Theorem 2.2 applies, there exist functions $A^{(1)}(G, T)$, $F_1^{(1)}(G, T)$ and $F_2^{(1)}(G, T)$ that are the unique solutions of the system $\Phi_A(G, A, F_1, F_2, T) = \Phi_{F_1}(A, F_1, F_2, T) = \Phi_{F_2}(F_2, T) = 0$ for each (G, T). These functions are differentiable on $\mathbb{R}_+^2$ by the implicit function theorem.

We then introduce $\Phi_T^{(1)}(G, T) = \Phi_T(G, A^{(1)}(G, T), F_1^{(1)}(G, T), F_2^{(1)}(G, T), T)$. The biological hypotheses imply that Theorem 2.1 applies and unicity in equations for A, F_1, F_2 implies that the function $\Phi_T^{(1)}(G, T)$ has a root in T for every G. We deduce that a sufficient condition for unicity is given by the Gale-Nikaido theorem applied on the model reduced to the variable G; more precisely, the function $\Phi_T^{(1)}$ is differentiable on $\mathbb{R}_+^2$. From the definition of the function $\Phi_T^{(1)}$ it follows $\frac{\partial \Phi_T^{(1)}}{\partial T} = \frac{\partial \Phi_T}{\partial T} + (\alpha_K\chi_A^{Krb} - \alpha_S\chi_A^{Sy})\frac{\partial A^{(1)}}{\partial T} + \alpha_{O1}\chi_{F_1}^{Ox1}\frac{\partial F_1^{(1)}}{\partial T} + (\alpha_{O1}R_{F_2}^{Ox1} + \alpha_{O2}R_{F_2}^{Ox2} + \alpha_S R_{F_2}^{Sy})\frac{\partial F_2^{(1)}}{\partial T}$.

We show by formal manipulation by using Mathematica version 5.2 software (in the derivation we neglect terms involving F_2, because of (4)) that the strong lipolytic conditions implies $\frac{\partial \Phi_T^{(1)}}{\partial T} < 0$. In other words, $\Phi_T^{(1)}$ is monotonic so that $\Phi_T^{(1)}(G, T)$ has a unique zero for every G. Let $T = T^{(2)}(G)$ be the solution of this equation. ∎

Biological interpretation. We now turn to the interpretation of the algebraic conditions. Although the systems of conditions has been reduced to a very abstract and condensed shape, exhibiting numerical coefficients to prove that the conditions are satisfied is not always possible. In order to do that we need either a (at least partially) parametrized model from the very beginning, or a series of experiments to estimate the control coefficients. As an alternative, let us express the algebraic conditions introduced in Theorem 4.1 as biological conditions over the relative strengths of fluxes and their dependencies on the products of the system.

- The <u>stoichiometry condition</u> (5) can be checked from biochemical data, by considering the average numbers of Acetyl-coA and ATP molecules produced or consumed by the different fluxes.

- The strong lipolytic condition (3) means that the energy variation has a sufficiently strong effect on the arrival of fatty acids inside the cell. Lee et al. [La04] studied for wild-type and PPAR-/- mutant murine liver, the fatty acids profiles in triglycerides (TG), which are the predominant ($> 50\%$) hepatic fatty acids and also in phospholipids (PL) which go into cellular membranes. Let us recall that TG and PL are storage forms of fatty acids and that PL contribute much less than TG to the total fatty acid mass. These authors [La04] show that for wild type hepatocytes after 72h of fasting fatty acids profiles do not change significantly in PL, but there is a strong increase of TG and of their fatty acids constituents. This suggests that the strong lipolytic condition is satisfied. To certify that the condition is always satisfied, we would nevertheless need experiments for every T constrained states as well.
- The fatty acid proportion condition (4) means that polyunsaturated fatty acids are minority among all FA, which is stated for instance in [La04].

It follows from this discussion that the conditions of uniqueness are reasonable for the biological viewpoint, which suggests that fatty acid metabolism and their regulation correspond to a model with a unique steady state instead of a bistable one.

4.3   Predictions of the Model and Some Validations

(a) **Role of genetic regulations in energy homeostasis.** Let $T = \mathrm{T}^{(2)}(\mathrm{G})$ be the unique solution of the equation $\Phi_\mathrm{T}^{(1)}(\mathrm{G}, \mathrm{T}) = 0$. The derivative $\frac{d\mathrm{T}^{(2)}}{d\mathrm{G}}$ is the appropriate quantity to investigate the role of genetic regulation in energy homeostasis. It quantifies the energy buffering effect: the lower is this derivative, hence the lower is the variation of T for a fixed variation of G, the stronger is the energy buffering. We use formal studies of signs to compare the values of $\frac{d\mathrm{T}^{(2)}}{d\mathrm{G}}$ at quasistationarity (steady state of genetically non-regulated model) and at stationarity (steady state of the full model). In order to formulate the next result let us denote by B_{eq}, B_{qs} the values at stationarity and at quasistationarity of the combination of control coefficients B defined in (2).

Proposition 4.1. *Assume that the strong lipolytic condition (3) and the stoichiometry condition (5) are satisfied. Assume that* $\left(R_\mathrm{T}^{\mathrm{In}2} - R_\mathrm{T}^{\mathrm{O}\times 2}\right)_{eq,qs} > 0$ *and* $B_{eq} > B_{qs}$. *Then* $\left(\frac{d\mathrm{T}^{(2)}}{d\mathrm{G}}\right)_{qs} > \left(\frac{d\mathrm{T}^{(2)}}{d\mathrm{G}}\right)_{eq} > 0$.

The proof of this property is a sign study performed with Mathematica (see Additional material in the Appendix). Notice that this proposition is strongly related to the steady state condition introduced in Theorem 4.1 since both assume the strong lipolytic and stoichiometry conditions. As discussed in the previous section, both hypothesis are biologically reasonable.

Comment on the conditions

- By using derivative computations, we obtain that $\frac{dF_2^{(2)}}{dG}$ have the same sign as $R_{\mathrm{T}}^{\mathrm{Ox2}} - R_{\mathrm{T}}^{\mathrm{In2}}$. Then, the condition $\left(R_{\mathrm{T}}^{\mathrm{In2}} - R_{\mathrm{T}}^{\mathrm{Ox2}}\right)_{eq,qs} > 0$ means that PUFAs increase during fasting and decrease during feeding, which is confirmed by experiments of Lee et al. [La04] on wild-type and PPAR-/- mutant murine liver. This suggests that this hypothesis is biologically reasonable.
- Additionally, the condition $B_{eq} > B_{qs}$ is equivalent to

$$B_1 \left(R_{\mathrm{F}_2}^{\mathrm{Ox1}}\right)_{eq} + B_2 \left(R_{\mathrm{F}_2}^{\mathrm{Ko}}\right)_{eq} + B_3 \left(R_{\mathrm{F}_2}^{\mathrm{Sy}}\right)_{eq} + B_4 \frac{\partial \mathrm{Ox2}}{\partial E3} \frac{\partial E3}{\partial \mathrm{F}_2} > 0,$$

with $B_1 > 0$, $B_4 \frac{\partial \mathrm{Ox2}}{\partial E3} \frac{\partial E3}{\partial \mathrm{F}_2} > 0$. This means that even if B_2, B_3 are negative the oxidation control term is strong enough to win. At fasting, this is a plausible supposition. We consequently deduce the following:

Biological prediction. <u>Genetic regulation reinforces the energy buffering effect: variations of ATP for a fixed variation of nutriments are less important when genetic regulations exist.</u>

(b) Effect of genetic perturbation Let us consider now the effect of PPAR knock-out on the model. Without PPAR, there is no longer a genetic control on oxidation, therefore we expect to have less energy buffering on fasting. Less obvious is what happens to the concentration of PUFA. Let $F_2^{(2)}(G)$ be the value of PUFA concentration as a function of G. Also, let $B_{WT,eq}$, $B_{PPAR-/-,eq}$ be the values at steady state in wild type and mutants of the coefficient B defined in (2).

Proposition 4.2. *Assume that the strong lipolytic condition (3) and the stoichiometry condition (5) are satisfied. Assume also that $\left(R_{\mathrm{T}}^{\mathrm{In2}} - R_{\mathrm{T}}^{\mathrm{Ox2}}\right)_{eq,qs} > 0$ and $B_{WT,eq} > B_{PPAR-/-,eq}$ then*

$$\left(\frac{dT^{(2)}}{dG}\right)_{eq,PPAR-/-} > \left(\frac{dT^{(2)}}{dG}\right)_{eq,WT}, \quad and \quad \left|\frac{dF_2^{(2)}}{dG}\right|_{eq,PPAR-/-} > \left|\frac{dF_2^{(2)}}{dG}\right|_{eq,WT}.$$

As before, this can be checked by symbolic manipulations (see <u>Appendix</u>). Biologically, the condition $B_{WT,eq} > B_{PPAR-/-,eq}$ is equivalent to $B_1 \left(R_{\mathrm{F}_2}^{\mathrm{Ox1}}\right)_{WT,eq} + B_2 \left(R_{\mathrm{F}_2}^{\mathrm{Ko}}\right)_{WT,eq} + B_4 \frac{\partial \mathrm{Ox2}}{\partial E3} \left(\frac{\partial E3}{\partial \mathrm{F}_2}\right)_{WT,eq} > 0$, with $B_1 > 0$, $B_4 \frac{\partial \mathrm{Ox2}}{\partial E3} \left(\frac{\partial E3}{\partial \mathrm{F}_2}\right)_{WT,eq} > 0$. This means that even if B_2 is negative the oxidation genetic control term is large enough to compensate. It follows:

Biological prediction [PPAR -/- mutants] <u>(a) PPAR knock-out reduces energy buffering. (b) The increase of PUFA concentration under fasting is stronger in PPAR knocked-out cells compared to the same increase in wild type cells.</u>

Experiments on transgenic mice showed that after a 72h-fast, fatty acids concentration increases at a higher extent in PPAR knocked-out cells with respect to wild type cells [BLC+04, BGG+09]. This is coherent with the observations by Lee et al.[La04] that for the same length of fasting time the hepatic accumulation of triacylglycerol is 2.8 fold higher in PPAR knocked-out than in wild-type mice. Hence, the global behavior of fatty acids is consistent with our predictions.

(c) **Dynamical predictions.** This model also allows to deduce results on the behavior of several metabolites. For instance, we have $\frac{dT^{(2)}}{dG} > 0$, meaning that ATP decreases at fasting (which is not a surprise). Moreover, $\left|\frac{dF_2^{(2)}}{dG}\right|_{qs} > \left|\frac{dF_2^{(2)}}{dG}\right|_{eq}$, meaning that the curves representing PUFA concentration during fasting must show an overshoot if the input of glucose is a discontinuous step-like decrease: in this case, the model predicts that the increase in PUFA concentration is greater immediately at quasi-stationarity than later at stationarity.

5   Discussion

We have proposed a methodology to build small complexity abstractions that integrate various qualitative aspects of regulated metabolism. As main feature, such abstractions are integrative (main processes together with their various regulation), low complexity abstraction, not dependant on specific numerical values, and allow to distinguish between quick metabolic and slow genetic response.

Our model copes with the main experimental findings on the behavior of regulated fatty acid metabolism in hepatocytes. Under fasting, the model shifts from a synthesis dominated regime to an oxidation/lipolysis dominated regime. This shift stabilizes energy, replacing food supply by reserve consumption. At short times, the shift is performed by metabolic control of synthesis, lipolysis and oxidation. At longer times, the regulatory effect of an increase of intracellular PUFA on the nuclear receptors PPAR and LXR reinforces this control. Refeeding shifts the system in the opposite direction. The catabolic part of this model has been, after exploding some lumped details, successfully used for quantitative predictions on the behavior of fatty acids pools and of the genetic regulation in murine models [BGG+09].

In this paper we have detailed how, using only sign constraints on partial derivatives of elementary fluxes, it is possible to check the possibility of observed properties of the system and to predict others.

Additionally, we have illustrated how this approach allows to reduce complex models into simpler models that have exactly the same steady states in terms of the remaining variables. In the process of reduction we compute symbolically the control coefficients of the reduced model from the derivatives of the elementary fluxes in the full model. The resulting expressions can be used for direct biological predictions as we did in this case study. Finally, full sequences of box equilibrations can provide conditions for uniqueness of the steady state.

Although the full procedure is not yet automatic, it could be done so in the future. An important algorithmic aim is to develop effective algorithms to test the Gale-Nikaido condition from determinant signs or from topological derived conditions. As suggested in Section 2.4, an important step in such algorithms would be to identify and eliminate from the model the non-essential loops which have vanishing contribution to multi-stationarity. Such a method, combining topological criteria and model reduction will be presented elsewhere.

Another problem to solve is the increasing complexity of the expressions of the control coefficients resulting from the reduction. A solution to keep this complexity within fixed bounds has been proposed in [RGZL08] in connection with numerical solutions of quasi-stationarity equations, but similar methods could be applied to symbolic calculations. The idea is to take into account the orders of magnitude of various quantities (say, control coefficients) and to use consistent asymptotic calculations allowing to identify the dominant terms in the solutions of partial steady state equations.

References

[BGG+09]   Blavy, P., Gondret, F., Guillou, H., Lagarrigue, S., Martin, P.G.P., van Milgen, J., Radulescu, O., Siegel, A.: A minimal model for hepatic fatty acid balance during fasting: Application to PPAR alpha-deficient mice. Journal of Theoretical Biology 261(2), 266–278 (2009)

[BLC+04]   Barnouin, S., Lassere, F., Cantiello, M., Guillou, H., Pineau, T., Martin, P.: A kinetic view of coordinate modulations of gene expression and metabolism in wild-type and PPARα-/- mice during fasting. In: Conference of the Paul Hamel Institute, Monaco (2004)

[CB95]   Cornish-Bowden, A.: Fundamentals of Enzyme Kinetics. Portland Press, London (1995)

[CF03]   Clément, K., Ferré, P.: Genetics and the pathophysiology of obesity. Pediatric Research 53, 721–725 (2003)

[CTF06]   Craciun, G., Tang, Y., Feinberg, M.: Understanding bistability in complex enzyme-driven reaction networks. Proceedings of the National Academy of Sciences 103(23), 8697 (2006)

[DF04]   Duplus, E., Forest, C.: Is there a single mechanism for fatty acid regulation of gene transcription? Biochemical Pharmacology 64, 893–901 (2004)

[Fel97]   Fell, D.: Understanding the control of metabolism. Portland Press, London (1997)

[Jum04]   Jump, D.B.: Fatty acid regulation of gene transcription. Critical Rev. in Clinical Lab. Sci. 41, 41–78 (2004)

[KKS+06]   Kitayama, T., Kinoshita, A., Sugimoto, M., Nakayama, Y., Tomita, M.: A simplified method for power-law modelling of metabolic pathways from time-course data and steady-state flux profiles. Theoretical Biology and Medical Modelling 3, 24 (2006)

[La04]   Lee, S., et al.: Requirement of PPARα in maintaining phospolipid and triaglycerol homeostasis during energy deprivation. J. Lipid Res. 4, 2025–2037 (2004)

[LGP06]   Lee, J.M., Gianchandani, E.P., Papin, J.A.: Flux balance analysis in the era of metabolomics. Briefings in Bioinformatics 7(2), 140 (2006)

[Par83]   Parthasarathy, T.: On Global Univalence Theorems. Lecture Notes in Mathematics, vol. 977. Springer, Heidelberg (1983)

[PLM03]   Pégorier, J.-P., Le May, C.: Régulation de l'expression génique par les acides gras (Control of gene expression by fatty acids). Nutrition clinique et métabolisme 17, 80–88 (2003)

[RGZL08]   Radulescu, O., Gorban, A.N., Zinovyev, A., Lilienbaum, A.: Robust
           simplifications of multiscale biochemical networks. BMC Systems Biol-
           ogy 2(1), 86 (2008)
[RLS+06]   Radulescu, O., Lagarrigue, S., Siegel, A., Veber, P., Le Borgne, M.:
           Topology and linear response of interaction networks in molecular biol-
           ogy. Journal of The Royal Society Interface 3(6), 185–196 (2006)
[SV87]   Savageau, M.A., Voit, E.O.: Recasting nonlinear differential equations as
           S-systems: a canonical nonlinear form. Mathematical biosciences 87(1),
           83–115 (1987)
[Tho81]   Thomas, R.: On the relation between the logical structure of systems and
           their ability to generate multiple steadt states or sustained oscillations.
           Springer Ser. Synergetics 9, 180–193 (1981)

A   Appendix: Additional Detailed Proofs

A.1   Proof of Theorem 2.1

Proof of Lemma 2.1 By the Poincaré-Hopf formula a sufficient condition for having a zero in the interior of D is to have a non-zero index for the vector field on S. Since Φ points inward D on S, we can construct a smooth change of variables which conjugates Φ on a neighborhood of D to a vector field Φ' defined on a neighborhood of the unit $n-$ball $\mathbf{B}^n$, such that on a neighborhood of the unit $n-$sphere $\mathbf{S}^n$, Φ' coincides with the radial vector field $\mathbf{X} \mapsto -\mathbf{X}$. For this vector field Φ', we can compute its index, which is 1 or -1 according to the parity of n. The Lemma is proved since the index is a differential invariant.   ∎

Proof of Theorem 2.1. From Lemma 2.1, it is enough to find a smooth ball in the positive orthant on the boundary of which the vector field Φ points inwards.

For $R > 0$, let us consider the intersection domain of the closed n-ball of radius R with the positive orthant: $\Delta = \{\mathbf{X} \in \mathbb{R}^n_+, \|\mathbf{X}\| \leq R\}$. This domain is a topological ball; let us denote Σ its boundary. If $\mathbf{X} \in \Sigma$ and none of its components is 0, then for R large enough, $\Phi(\mathbf{X})$ points inward Δ, because G is bounded and $\Lambda_i(\mathbf{X})$ tend to infinity with $\mathbf{X}$, hence $\Phi_i(\mathbf{X}) < 0$, for all $1 \leq i \leq n$. On the other hand, if only one of the components of $\mathbf{X}$ is 0, then by hypothesis (2), $\Phi(\mathbf{X})$ points inward Δ. Since the set of points where the property of pointing inwards is open, we can find a smooth ball D contained in Δ and sufficiently close to it, such that on the boundary of D, the Φ points inward D.   ∎

A.2   Computational Details for Theorem 4.1

The derivatives $\frac{\partial A^{(1)}}{\partial T}, \frac{\partial F_1^{(1)}}{\partial T}, \frac{\partial F_2^{(1)}}{\partial T}$ are obtained as

$$\frac{\partial}{\partial T}\begin{pmatrix} A^{(1)} \\ F_1^{(1)} \\ F_2^{(1)} \end{pmatrix} = -(J^{(1)})^{-1}\frac{\partial}{\partial T}\begin{pmatrix} \Phi_A \\ \Phi_{F_1} \\ \Phi_{F_2} \end{pmatrix}.$$

They thus can be expressed by means of fluxes and of control coefficients in the following way:

$$-\frac{\det(J^{(1)})}{\chi_{F_1}^{tot}}\frac{\partial A^{(1)}}{\partial T} = \chi_{F_2}^{tot}\{R_T^{Gly} - R_T^{Krb} + n_2[\rho_{F_2}^{Ox2}R_T^{In2} + (1 - \rho_{F_2}^{Ox2})R_T^{Ox2}] +$$

$$+n_1[(1 - \rho_{F_1}^{Ox1})(R_T^{Ox1} + R_T^{Sy})\rho_{F_1}^{Ox1}R_T^{In1}]\} + (R_T^{In2} - R_T^{Ox2})[-R_{F_2}^{Ko}$$

$$+n_1(R_{F_2}^{Ox1} + R_{F_2}^{Sy})(1 - \rho_{F_1}^{Ox1})]$$

$$-\frac{\det(J^{(1)})}{\chi_A^{tot}}\frac{\partial F_1^{(1)}}{\partial T} = \chi_{F_2}^{tot}\{\rho_A^{Sy}(R_T^{Gly} - R_T^{Krb} + n_2 R_T^{Ox2}) + n_1[R_T^{In1} - (1 - \rho_A^{Sy})(R_T^{Ox1}$$

$$+R_T^{Sy})]\} + (R_T^{Ox2} - R_T^{In2})[n_1(R_{F_2}^{Ox1} + R_{F_2}^{Sy}) - n_2 R_{F_2}^{Ox2} + \rho_A^{Sy}R_{F_2}^{Ko}]$$

$$\frac{\partial F_2^{(1)}}{\partial T} = (\chi_{F_2}^{tot})^{-1}(R_T^{Ox2} - R_T^{In2}) \tag{6}$$

where $-\det(J^{(1)}) = \chi_{F_2}^{tot}\chi_A^{tot}\chi_{F_1}^{tot}(1 - \rho_A^{Sy}\rho_{F_1}^{Ox1})$. We deduce $\frac{\partial \Phi_T^{(1)}}{\partial T} = [A(R_T^{Ox1} - R_T^{In1}) + B(R_T^{Ox2} - R_T^{In2}) + C - D]/[n_1(1 - \rho_A^{Sy}\rho_{F_1}^{Ox1})]$ where A, B, C, D are combinations of control parameters defined in Eq. (2).

We also prove that if the stoichiometry condition (5) is fulfilled, then $X > 0, A > 0, B_1 > 0, B_4 > 0, C > 0, D > 0$ with a lengthy but straightforward formal manipulation of (6). We have gathered control coefficients into as large as possible positive combinations. As an illustration of how the stoichiometry condition was used let us consider the sign of D. From $\alpha_{O1} > \alpha_S$, $X/n_1 - \alpha_{O1}(1 - \rho_A^{Sy}) = (\alpha_{O1} - \alpha_S)\rho_A^{Sy} + n_1\alpha_K\rho_A^{Krebs} > 0$, $\rho_A^{Sy} < 1$ and $\rho_{F_1}^{Ox1} < 1$ it follows that $D > 0$.

A.3   Proof of Propositions 4.1 and 4.2

Lemma A.1. *Let* $F_2^{(2)}(G) = F_2^{(1)}(G, T^{(2)}(G))$. *If the strong lipolytic condition is satisfied, then the sign of* $\frac{dF_2^{(2)}}{dG}$ *is equal to the sign of* $R_T^{Ox2} - R_T^{In2}$.

Proof. The chain rule gives

$$\frac{dF_2^{(2)}}{dG} = \frac{\partial F_2^{(1)}}{\partial G} + \frac{\partial F_2^{(1)}}{\partial T}\frac{dT^{(2)}}{dG}.$$

Since $\frac{\partial}{\partial G}\begin{pmatrix} A^{(1)} \\ F_1^{(1)} \\ F_2^{(1)} \end{pmatrix} = -(J^{(2)})^{-1}\begin{pmatrix} R_G^{Gly} \\ 0 \\ 0 \end{pmatrix} = \frac{R_G^{Gly}}{\chi_A^{tot}\chi_{F_1}^{tot}(1 - \rho_A^{Sy}\rho_{F_1}^{Ox1})}\begin{pmatrix} \chi_{F_1}^{tot} \\ \chi_A^{Sy} \\ 0 \end{pmatrix}$ we have

$\frac{\partial F_2^{(1)}}{\partial G} = 0$. It follows from Eq. (6) – computations related to the proof of Theorem 4.1 – that the sign of $\frac{\partial F_2^{(1)}}{\partial T}$ is the same as the sign of $R_T^{Ox2} - R_T^{In2}$. Moreover, if the strong lipolytic condition and the stoichiometry condition are satisfied, then $\frac{dT^{(2)}}{dG} > 0$.   ■

Proof of Proposition 4.1. The differences between stationarity and quasi-stationarity occur at two levels:

1. At quasi-stationarity F_2 does not regulate the genetic variables:

$$\left(R_{F_2}^{Sy}\right)_{qs} = \left(R_{F_2}^{Ox1}\right)_{qs} = \left(R_{F_2}^{Ko}\right)_{qs} = 0. \tag{7}$$

2. At quasi-stationarity the control of F_2 on its oxidation is only a metabolic substrate effect. Genetic control is added at stationarity. We have $R_{F_2}^{Ox2} = \frac{\partial Ox2}{\partial F_2} + \frac{\partial E_3}{\partial F_2}\frac{\partial Ox2}{\partial E_3}$ with $\frac{\partial Ox2}{\partial E_3} > 0$, and $\chi_{F_2}^{tot} = R_{F_2}^{Ox2} - \frac{\partial In2}{\partial F_2}$, $\rho_{F_2}^{Ox2} = (1 - \frac{\frac{\partial In2}{\partial F_2}}{R_{F_2}^{Ox2}})^{-1}$, with $\frac{\partial In2}{\partial F_2} < 0$. Furthermore, $\left(\frac{\partial E_3}{\partial F_2}\right)_{eq} > 0$ and $\left(\frac{\partial E_3}{\partial F_2}\right)_{qs} = 0$. Hence:

$$\left(R_{F_2}^{Ox2}\right)_{eq} > \left(R_{F_2}^{Ox2}\right)_{qs}, \quad \left(\chi_{F_2}^{tot}\right)_{eq} > \left(\chi_{F_2}^{tot}\right)_{qs}, \quad \left(\rho_{F_2}^{Ox2}\right)_{eq} > \left(\rho_{F_2}^{Ox2}\right)_{qs}. \tag{8}$$

We easily compute the following

$$\frac{\partial \Phi_T^{(1)}}{\partial G} = \frac{R_G^{Gly}\{n_1\alpha_K\chi_A^{Krb} + \chi_A^{tot}[n_1\alpha_G(1-\rho_{F_1}^{Ox1}\rho_A^{Sy}) + \alpha_{O1}\rho_{F_1}^{Ox1}\rho_A^{Sy} - \alpha_S\rho_A^{Sy}]\}}{n_1\chi_A^{tot}(1-\rho_A^{Sy}\rho_{F_1}^{Ox1})}$$

We deduce that $\frac{\partial \Phi_T^{(1)}}{\partial G}$ is the same at stationarity and at quasi-stationarity. From Theorem 4.1, it follows :

$$\frac{\partial \Phi_T^{(1)}}{\partial T} = \mathcal{R} - \frac{B}{n_1(1-\rho_A^{Sy}\rho_{F_1}^{Ox1})}(R_T^{In2} - R_T^{Ox2}), \tag{9}$$

where $\mathcal{R}$ is a term not changing from quasi-stationarity to stationarity and the expression of B is given in (2).

It remains to notice that $\frac{dT^{(2)}}{dG} = -\left(\frac{\partial \Phi_T^{(1)}}{\partial G}\right)/\left(\frac{\partial \Phi_T^{(2)}}{\partial T}\right)$ and $B_{qs} < B_{eq}$ to conclude $\left(\frac{dT^{(2)}}{dG}\right)_{qs} > \left(\frac{dT^{(2)}}{dG}\right)_{eq}$. From $\frac{dF_2^{(2)}}{dG} = -\frac{dT^{(2)}}{dG}(R_T^{In2} - R_T^{Ox2})/\chi_{F_2}^{tot}$ it also follows that $\left|\frac{dF_2^{(2)}}{dG}\right|_{qs} > \left|\frac{dF_2^{(2)}}{dG}\right|_{eq}$. ∎

Proof of Proposition 4.2. We follows closely the proof of Prop. 4.1. The differences between $PPAR-/-$ and WT cells occur at two levels:

$$\left(R_{F_2}^{Ox1}\right)_{PPAR-/-} = \left(R_{F_2}^{Ko}\right)_{PPAR-/-} = 0 \tag{10}$$

$$\left(R_{F_2}^{Ox2}\right)_{WT,eq} > \left(R_{F_2}^{Ox2}\right)_{PPAR-/-,eq}, \quad \left(\chi_{F_2}^{tot}\right)_{WT,eq} > \left(\chi_{F_2}^{tot}\right)_{PPAR-/-,eq},$$

$$\left(\rho_{F_2}^{Ox2}\right)_{WT,eq} > \left(\rho_{F_2}^{Ox2}\right)_{PPAR-/-,eq} \tag{11}$$

If $B_{WT,eq} > B_{PPAR-/-,eq}$, it follows (along the same lines as the proof of Prop. 4.1) that

$\left(\frac{dT^{(2)}}{dG}\right)_{PPAR-/-,eq} > \left(\frac{dT^{(2)}}{dG}\right)_{WT,eq}$. From $\frac{dF_2^{(2)}}{dG} = -\frac{dT^{(2)}}{dG}(R_T^{In2} - R_T^{Ox2})/\chi_{F_2}^{tot}$ and Eq.(11) it follows that $\left|\frac{dF_2^{(2)}}{dG}\right|_{eq,PPAR-/-} > \left|\frac{dF_2^{(2)}}{dG}\right|_{eq,WT}$. ∎

Cultural Epigenetics: On the Heritability of Complex Diseases

Rodrick Wallace[1,*] and Deborah Wallace[2]

[1] The New York State Psychiatric Institute,
549 W. 123 St., 16F, New York, NY, 10027
Tel.: 212-865-4766
wallace@pi.cpmc.columbia.edu
[2] Consumers Union
rdwall@ix.netcom.com

Abstract. We extend a cognitive paradigm for gene expression based on the asymptotic limit theorems of information theory to the epigenetic epidemiology of complex developmental disorders in humans. In particular, we recognize the fundamental role culture plays in human biology, a heritage mechanism parallel to, and interacting with, the more familiar genetic and epigenetic systems. We do this via a model through which culture acts as another tunable epigenetic catalyst that both directs developmental trajectories, and becomes convoluted with individual ontology, via a mutually-interacting crosstalk mediated by a social interaction that is itself culturally driven. In sum, embedding culture is an essential component of the regulation of human development and its dysfunctions. The cultural and epigenetic systems of heritage may thus provide the 'missing' heritability of complex diseases that is currently the subject of much scientific discourse.

1 Introduction

1.1 Mental Disorders and Culture

We begin with a discussion of human mental disorders, that, while increasingly recognized as quintessentially developmental, remain deeply mysterious. Our classic scientific task will be to infer the general from the particular, extending our focus on the central role of culture in human mental dysfunction to a vastly larger spectrum of developmental pathologies, with the hope of shedding some light on the missing heritability conundrum [66]. This task, it seems, requires cutting-edge mathematical methods for even a basic formal analysis.

The understanding of mental disorders is in considerable disarray. Official classifications of mental illness such as the *Diagnostic and Statistical Manual of Mental Disorders - Fourth Edition* [34], the standard descriptive nosology in the US, have even been characterized as 'prescientific' by P. Gilbert [42] and others. Johnson-Laird et al. [60] claim that current knowledge about psychological

---

* Corresponding author.

C. Priami et al. (Eds.): Trans. on Comput. Syst. Biol. XIII, LNBI 6575, pp. 131–170, 2011.
© Springer-Verlag Berlin Heidelberg 2011

illnesses is comparable to the medical understanding of epidemics in the early 19th century. Physicians realized then that cholera, for example, was a specific disease, which killed about a third of the people whom it infected. What they disagreed about was the cause, the pathology, and the communication of the disease. Similarly, according to [60], most medical professionals these days realize that psychological illnesses occur (cf. [34]), but they disagree about their cause and pathology. Notwithstanding [34], Johnson-Laird et al. doubt whether any satisfactory a priori definition of psychological illness can exist because it is a matter for theory to elucidate.

Atmanspacher [4], however, argues that formal theory of high level cognitive process is itself in a disarray similar to that of physics 400 years ago, in that the basic entities, and the relations between them, have yet to be delineated.

More generally, simple arguments from genetic determinism regarding mental and other developmental disorders fail, in part because of a draconian population bottleneck that, early in our species' history, resulted in an overall genetic diversity less than that observed within and between contemporary chimpanzee subgroups. Manolio et al. [66], in a major review article, express this as 'finding the missing heritability of complex diseases'. They observe, for example, that at least 40 loci have been associated with human height, a classic complex trait with an estimated heritability of about 80 %, yet they explain only about 5 % of phenotype variance despite studies of tens of thousands of people. This result, they find, is typical across a broad range of supposedly heritable diseases, and call for extending beyond current genome-wide assoication approaches to illuminate the genetics of complex diseases and enhance its potential to enable effective disease prevention or treatment.

Arguments from psychosocial stress fare better (e.g., [14, 31, 35]), particularly for depression, but are affected by the apparently complex and contingent developmental paths determining the onset of schizophrenia, dementias, psychoses, and so forth, some of which may be triggered in utero by exposure to infection, low birthweight, or other functional teratogens.

P. Gilbert [42] suggests an extended evolutionary perspective, in which evolved mechanisms like the 'flight-or-fight' response are inappropriately excited or suppressed, resulting in such conditions as anxiety or post traumatic stress disorders. Nesse [75] suggests that depression may represent the dysfunction of an evolutionary adaptation which down-regulates foraging activity in the face of unattainable goals.

Kleinman and Good, however, ([63], p. 492) outline something of the cross cultural subtleties affecting the study of depression that seem to argue against any simple evolutionary or genetic interpretation. They state that, when culture is treated as a constant, as is common when studies are conducted in our own society, it is relatively easy to view depression as a biological disorder, triggered by social stressors in the presence of ineffective support, and reflected in a set of symptoms or complaints that map back onto the biological substrate of the disorder. However, they continue, when culture is treated as a significant variable, for example, when the researcher seriously confronts the world of meaning and experience of members of non-Western societies, many of our assumptions about the nature of emotions and illness are cast in sharp relief. Dramatic differences are found across cultures in the social organization, personal experience, and consequences of such emotions as sadness, grief, and anger, of behaviors such as withdrawal or aggression, and of psychological characteristics such as passivity and helplessness or the resort to altered states of consciousness. They are organized differently as psychological realities, communicated in a wide range of idioms, related to quite varied local contexts of power relations, and are interpreted, evaluated, and responded to as fundamentally different meaningful realities. Depressive illness and dysphoria are thus not only interpreted differently in non-Western societies and across cultures; they are *constituted* as fundamentally different forms of social reality.

Since publication of that landmark study, a number of comprehensive overviews have been published that support its conclusions [8, 59, 67]. As Marsella [64] writes, it is now clear that cultural variations exist in the areas of meaning, perceived causes, onset patterns, epidemiology, symptom expression, course, and outcome, variations having important implications for understanding clinical activities including conceptualization, assessment, and therapy.

Kleinman and Cohen [64] argue forcefully that several myths have become central to Western psychiatry. The first is that the forms of mental illness everywhere display similar degrees of prevalence. The second is an excessive adherence to a principle known as the pathogenic/pathoplastic dichotomy, which holds that biology is responsible for the underlying structure of a malaise, whereas cultural beliefs shape the specific ways in which a person experiences it. The third myth maintains that various unusual culture-specific disorders whose biological bases are uncertain occur only in exotic places outside the West. In an effort to base psychiatry in 'hard' science and thus raise its status to that of other medical disciplines, psychiatrists have narrowly focused on the biological underpinnings of mental disorders while discounting the importance of such 'soft' variables as culture and socioeconomic status.

Heine [49] describes an explicit cultural psychology that views the person as containing a set of biological potentials interacting with particular situational contexts that constrain and afford the expression of various constellations of traits and patterns of behavior. He says that, unlike much of personality psychology, cultural psychology focuses on the constraints and affordances inherent to the cultural environment that give shape to those biological potentials. Human nature, from this perspective, is seen as emerging from participation in cultural worlds, and of adapting oneself to the imperatives of cultural directives, meaning that our nature is ultimately that of a cultural being.

Heine describes how cultural psychology does not view culture as a superficial wrapping of the self, or as a framework within which selves interact, but as something that is intrinsic to the self, so that without culture there is no self, only a biological entity deprived of its potential. Individual selves, from Heine's perspective, are inextricably grounded in a configuration of consensual understandings and behavioral customs particular to a given cultural and historical context, so that understanding the self requires an understanding of the culture that sustains it. Heine argues, then, that the process of becoming a self is contingent on individuals interacting with, and seizing meanings from, the cultural environment.

Heine warns that the extreme nature of American individualism means that a psychology based on late 20th century American research not only stands the risk of developing models that are particular to that culture, but of developing an understanding of the self that is peculiar in the context of the world's cultures.

Indeed, as Norenzayan and Heine [77] point out, for the better part of a century, a considerable controversy has raged within anthropology regarding the degree to which psychological and other human universals do, in fact, actually exist independent of the particularities of culture.

Many others have made similar points over the years (e.g., [1, 51, 68, 69, 76, 96]).

As Durham [33] and Richerson and Boyd [82] explore at some length, humans are endowed with two distinct but interacting heritage systems: genes and culture. Durham [33], for example, writes that genes and culture constitute two distinct but interacting systems of information inheritance within human populations and information of both kinds has influence, actual or potential, over behaviors, which creates a real and unambiguous symmetry between genes and phenotypes on the one hand, and culture and phenotypes on the other. Genes and culture, in his view, are best represented as two parallel lines or tracks of hereditary influence on phenotypes.

Both genes and culture can be envisioned as generalized languages in that they have recognizable 'grammar' and 'syntax', in the sense of [78, 95, 100, 101].

More recent work has identified epigenetic heritage mechanisms involving such processes as environmentally-induced gene methylation, that can have strong influence across several generations (e.g., [54-6, 101]), and are the subject of intense current research, a matter to which we will return below.

There are, it seems, two powerful heritage mechanisms in addition to the genetic where one may perhaps find the 'missing heritability of complex diseases' that Manolio et al. [66] seek.

Here we will expand recent explorations of a cognitive paradigm for gene expression [100, 101] that incorporates the effects of surrounding epigenetic regulatory machinery as a kind of catalyst to include the effects of the embedding information source of human culture on human ontology. The essential feature is that a cognitive process, including gene expression, can instantiate a dual information source that can interact with the generalized language of culture in which, for example, social interplay and the interpretation of socioeconomic and environmental stressors, involve complicated matters of symbolism and its grammar and syntax. These information sources interact by a crosstalk that, over the life course, determines human ontology and its manifold dysfunctions.

2   A Cognitive Paradigm for Gene Expression

As described in [100, 101], a cognitive paradigm for gene expression is under active study, a model in which contextual factors determine the behavior of what Cohen characterizes as a 'reactive system', not at all a deterministic – or even simple stochastic – mechanical process (e.g., [23, 24, 100, 101]). The various formal approaches are, however, all in the spirit of Maturana and Varela [70, 71] who understood the central role that cognitive process must play across a vast array of biological phenomena.

O'Nuallain [78] puts gene expression directly in the realm of complex linguistic behavior, for which context imposes meaning. He claims that the analogy between gene expression and language production is useful both as a fruitful research paradigm and also, given the relative lack of success of natural language processing (nlp) by computer, as a cautionary tale for molecular biology. A relatively simple model of cognitive process as an information source permits use of Dretske's [32] insight that any cognitive phenomenon must be constrained by the limit theorems of information theory, in the same sense that sums of stochastic variables are constrained by the Central Limit Theorem. This perspective permits a new formal approach to gene expression and its dysfunctions, in particular suggesting new and powerful statistical tools for data analysis that could contribute to exploring both ontology and its pathologies. Here we extend the mathematical foundations of that work to examine the topological structures of development and developmental disorder, in the context of an embedding information source representing the compelling varieties of human culture.

This approach is consistent with the broad context of epigenetics and epigenetic epidemiology. Jablonka and Lamb [54, 55], for example, argue that information can be transmitted from one generation to the next in ways other than through the base sequence of DNA. It can be transmitted through cultural and behavioral means in higher animals, and by epigenetic means in cell lineages. All of these transmission systems allow the inheritance of environmentally induced variation. Such Epigenetic Inheritance Systems are the memory systems that enable somatic cells of different phenotypes but identical genotypes to transmit their phenotypes to their descendants, even when the stimuli that originally induced these phenotypes are no longer present.

After some years of active research and debate, this epigenetic perspective has received much empirical confirmation (e.g., [6, 56, 57, 91]).

Foley et al. [39] argue that epimutation is estimated to be 100 times more frequent than genetic mutation and may occur randomly or in response to the environment. Periods of rapid cell division and epigenetic remodeling are likely to be most sensitive to stochastic or environmentally mediated epimutation. Disruption of epigenetic profile is a feature of most cancers and is speculated to play a role in the etiology of other complex diseases including asthma allergy, obesity, type 2 diabetes, coronary heart disease, autism spectrum disorders and bipolar disorders and schizophrenia.

Important work by Scherrer and Jost [86, 87] that is similar to the approach of this paper explicitly invokes information theory in their extension of the definition of the gene to include the local epigenetic machinery, a construct they term the 'genon'. Their central point is that coding information is not simply contained in the coded sequence, but is, in their terms, *provided by* the genon that accompanies it on the expression pathway and controls in which peptide it will end up. In their view the information that counts is not about the identity of a nucleotide or an amino acid derived from it, but about the relative frequency of the transcription and generation of a particular type of coding sequence that then contributes to the determination of the types and numbers of functional products derived from the DNA coding region under consideration.

The proper formal tools for understanding phenomena that 'provide' information – that are information sources – are the Rate Distortion Theorem and its zero error limit, the Shannon-McMillan Theorem.

3   Models of Development

The currently popular spinglass model of development (e.g., [20, 21]) assumes that N transcriptional regulators, are represented by their expression patterns

$$\mathbf{S}(t) = [S_1(t), ..., S_N(t)] \tag{1}$$

at some time t during a developmental or cell-biological process and in one cell or domain of an embryo. The transcriptional regulators influence each other's expression through cross-regulatory and autoregulatory interactions described by a matrix $w = (w_{ij})$. For nonzero elements, if $w_{ij} > 0$ the interaction is activating, if $w_{ij} < 0$ it is repressing. w represents, in this model, the regulatory genotype of the system, while the expression state $\mathbf{S}(t)$ is the phenotype. These regulatory interactions change the expression of the network $\mathbf{S}(t)$ as time progresses according to a difference equation

$$S_i(t + \Delta t) = \sigma[\sum_{j=1}^{N} w_{ij} S_j(t)], \tag{2}$$

where Δt is a constant and σ a sigmodial function whose value lies in the interval $(-1, 1)$. In the spinglass limit σ is the sign function, taking only the values ± 1.

The regulatory networks of interest here are those whose expression state begins from a prespecified initial state $\mathbf{S}(0)$ at time $t = 0$ and converge to a prespecified stable equilibrium state $\mathbf{S}_\infty$. Such networks are termed *viable* and must necessarily be a very small fraction of the total possible number of networks, since most do not begin and end on the specified states. This 'simple' observation is not at all simple in our reformulation, although other results become far more accessible, as we can then invoke the asymptotic limit theorems of information theory.

The spinglass approach to development is formally similar to spinglass neural network models of learning by selection, e.g., as proposed by Toulouse et al. [90] nearly a generation ago. Much subsequent work, summarized by Dehaene and Naccache [27], suggests that such models are simply not sufficient to the task of understanding high level cognitive function, and these have been largely supplanted by complicated 'global workspace' concepts whose mathematical characterization is highly nontrivial [4].

Wallace and Wallace [100, 101] shift the perspective on development by invoking a cognitive paradigm for gene expression, following the example of the Atlan/Cohen model of immune cognition.

Atlan and Cohen [3], in the context of a study of the immune system, argue that the essence of cognition is the comparison of a perceived signal with an internal (learned or inherited) picture of the world, and then, upon that comparison, the choice of a single response from a larger repertoire of possible responses.

Such choice inherently involves information and information transmission since it always generates a reduction in uncertainty, as explained by Ash ([2], p. 21).

More formally, a pattern of incoming input – like the $S(t)$ above – is mixed in a systematic algorithmic manner with a pattern of internal ongoing activity – like the (w_{ij}) above – to create a path of combined signals $x = (a_0, a_1, ..., a_n, ...)$ – analogous to the sequence of $S(t + \Delta t)$ above, with, say, $n = t/\Delta t$. Each a_k thus represents some functional composition of internal and external signals.

For a cognitive process, this path is supposed to be fed into a 'highly nonlinear decision oscillator', h, a sudden threshold machine whose canonical model could well be taken as the famous integrate-and-fire neuron (e.g., [53], Prop. 8.12). $h(x)$, otherwise seen as a 'black box', thus generates an output that is an element of one of two disjoint sets B_0 and B_1 of possible system responses. Let us define the sets B_k as

$$B_0 \equiv \{b_0, ..., b_k\},$$
$$B_1 \equiv \{b_{k+1}, ..., b_m\}. \tag{3}$$

Assume a graded response, supposing that if $h(x) \in B_0$, the pattern is not recognized, and if $h(x) \in B_1$, the pattern has been recognized, and some action $b_j, k + 1 \leq j \leq m$ takes place.

Rather than focusing on the properties of h, we *shift the perspective*: The principal objects of formal interest become *paths* x triggering pattern recognition-and-response. That is, given a fixed initial state a_0, examine all possible subsequent paths x beginning with a_0 and leading to the event $h(x) \in B_1$. Thus $h(a_0, ..., a_j) \in B_0$ for all $0 < j < m$, but $h(a_0, ..., a_m) \in B_1$.

Several points are central to the shift in perspective we are making:

(1). It is important to understand that the fundamental core of the argument does not regard the exact internal details of the inferred (but perhaps not easily observed) function $h(x)$, but rather has been shifted to the 'grammar' and 'syntax' of the strings $x = a_0, a_1, ...$ leading to action of that function, and which are more likely to be observable. We are concerned, then, with rules of operation rather than structural blueprints or de-facto circuit wirings, as interesting and important as these may be in other contexts.

(2). For each positive integer n, let $N(n)$ be the number of high probability grammatical and syntactical paths of length n that begin with some particular a_0 and lead to the condition $h(x) \in B_1$. Call such paths 'meaningful', assuming, not unreasonably, that $N(n)$ will be considerably less than the number of all possible paths of length n leading from a_0 to the condition $h(x) \in B_1$.

(3). While the combining algorithm, the form of the 'nonlinear oscillator' h, and the details of grammar and syntax are all unspecified in this model, *the critical assumption* that permits inference of the necessary conditions constrained by the asymptotic limit theorems of information theory is that the finite limit

$$H \equiv \lim_{n \to \infty} \frac{\log[N(n)]}{n} \qquad (4)$$

both exists and is independent of the path x.

Define such a pattern recognition-and-response cognitive process as *ergodic*. Not all cognitive processes are likely to be ergodic in this sense, implying that H, if it indeed exists at all, is path dependent, although extension to nearly ergodic processes seems possible [98].

Invoking the spirit of the Shannon-McMillan Theorem, as choice involves an inherent reduction in uncertainty, it is then possible to define an adiabatically, piecewise stationary, ergodic (APSE) information source $\mathbf{X}$ associated with stochastic variates X_j having joint and conditional probabilities $P(a_0, ..., a_n)$ and $P(a_n | a_0, ..., a_{n-1})$ such that appropriate conditional and joint Shannon uncertainties satisfy the classic relations

$$\begin{aligned} H[\mathbf{X}] &= \lim_{n \to \infty} \frac{\log[N(n)]}{n} \\ &= \lim_{n \to \infty} H(X_n | X_0, ..., X_{n-1}) \\ &= \lim_{n \to \infty} \frac{H(X_0, ..., X_n)}{n+1}. \end{aligned} \qquad (5)$$

This information source is defined as *dual* to the underlying ergodic cognitive process.

Adiabatic means that the source has been parametrized according to some scheme, and that, over a certain range, along a particular piece, as the parameters vary, the source remains as close to stationary and ergodic as needed for information theory's central theorems to apply. *Stationary* means that the system's probabilities do not change in time, and *ergodic*, roughly, that the cross sectional means approximate long-time averages. Between pieces it is necessary to invoke various kinds of phase transition formalisms, as described more fully in [95, 100].

In the developmental vernacular of [20, 21]., we now examine paths in phenotype space that begin at some $\mathbf{S}_0$ and converge $n = t/\Delta t \to \infty$ to some other $\mathbf{S}_\infty$. Suppose the system is conceived at $\mathbf{S}_0$, and h represents (for example) reproduction when phenotype $\mathbf{S}_\infty$ is reached. Thus $h(x)$ can have two values, i.e., B_0 not able to reproduce, and B_1, mature enough to reproduce. Then $x = (\mathbf{S}_0, \mathbf{S}_{\Delta t}, ..., \mathbf{S}_{n\Delta t}, ...)$ until $h(x) = B_1$.

Structure is now subsumed *within the sequential grammar and syntax of the dual information source* rather than within the cross sectional internals of (w_{ij})-space, a simplifying shift in perspective.

This transformation carries considerable computational burdens, as well as, and perhaps in consequence of, providing deep mathematical insight.

First, the fact that viable networks comprise a tiny fraction of all those possible emerges easily from the spinglass formulation simply because of the 'mechanical' limit that the number of paths from S_0 to S_∞ will always be far smaller than the total number of possible paths, most of which simply do not end on the target configuration.

From the information source perspective, which inherently subsumes a far larger set of dynamical structures than possible in a spinglass model – not simply those of symbolic dynamics – the result is what Khinchin [62] characterizes as the 'E-property' of a stationary, ergodic information source. This property allows, in the limit of infinitely long output, the classification of output strings into two sets:

(1). a very large collection of gibberish which does not conform to underlying (sequential) rules of grammar and syntax, in a large sense, and which has near-zero probability, and

(2). a relatively small 'meaningful' set, in conformity with underlying structural rules, having very high probability.

The essential content of the Shannon-McMillan Theorem is that, if $N(n)$ is the number of meaningful strings of length n, then the uncertainty of an information source X can be defined as

$$H[X] = \lim_{n \to \infty} \log[N(n)]/n,$$

that can be expressed in terms of joint and conditional probabilities. Proving these results for general stationary, ergodic information sources requires considerable mathematical machinery (e.g., [25, 29, 62]).

Second, according to [2], information source uncertainty has an important heuristic interpretation in that we may regard a portion of text in a particular language as being produced by an information source. A large uncertainty means, by the Shannon-McMillan Theorem, a large number of 'meaningful' sequences. Thus given two languages with uncertainties H_1 and H_2 respectively, if $H_1 > H_2$, then in the absence of noise it is easier to communicate in the first language; more can be said in the same amount of time. On the other hand, it will be easier to reconstruct a scrambled portion of text in the second language, since fewer of the possible sequences of length n are meaningful.

Third, information source uncertainty is homologous with free energy density in a physical system, a matter having implications across a broad class of dynamical behaviors.

The free energy density of a physical system having volume V and partition function $Z(K)$ derived from the system's Hamiltonian – the energy function – at inverse temperature K is (e.g., [65])

$$F[K] = \lim_{V \to \infty} -\frac{1}{K} \frac{\log[Z(K,V)]}{V}$$

$$= \lim_{V \to \infty} \frac{\log[\hat{Z}(K,V)]}{V}, \tag{6}$$

where $\hat{Z} = Z^{-1/K}$.

The partition function for a physical system is the normalizing sum in an equation having the form

$$P[E_i] = \frac{\exp[-E_i/kT]}{\sum_j \exp[-E_j/kT]} \tag{7}$$

where E_i is the energy of state i, k a constant, and T the system temperature.

Feynman [38], following the classic approach by Bennett [10], who examined idealized machines using information to do work, concludes that *the information contained in a message is most simply measured by the free energy needed to erase it.*

Thus, according to this argument, source uncertainty is homologous to free energy density as defined above, i.e., from the similarity with the relation $H = \lim_{n \to \infty} \log[N(n)]/n$.

Ash's perspective then has an important corollary: If, for a biological system, $H_1 > H_2$, source 1 will require more metabolic free energy than source 2.

4   Tunable Epigenetic Catalysis

Following [101], incorporating the influence of embedding contexts – generalized epigenetic effects – is most elegantly done by invoking the Joint Asymptotic Equipartition Theorem (JAEPT) [25]. For example, given an embedding epigenetic information source, say Y, that affects development, then the dual cognitive source uncertainty $H[X]$ is replaced by a joint uncertainty $H(X,Y)$. The objects of interest then become the jointly typical dual sequences $z^n = (x^n, y^n)$, where x is associated with cognitive gene expression and y with the embedding epigenetic regulatory context. Restricting consideration of x and y to those sequences that are in fact jointly typical allows use of the information transmitted from Y to X as the splitting criterion.

One important inference is that, from the information theory 'chain rule' [25], $H(X,Y) = H(X) + H(Y|X) \le H(X) + H(Y)$, while there are approximately $\exp[nH(X)]$ typical X sequences, and $\exp[nH(Y)]$ typical Y sequences, and hence $\exp[n(H(x) + H(Y))]$ independent joint sequences, there are only

$$\exp[nH(X,Y)] \le \exp[n(H(X) + H(Y))]$$

jointly typical sequences. Equality occurs only for stochastically independent processes.

Interpreting the homology between information and free energy rather broadly – making something of a an intuitive leap – the embedding context can be said to lower an analog to the activation energy of a particular developmental channel, at the expense of raising the total free energy needed, since the system must now support two information sources instead of one, i.e., that regulated, and that providing the regulation.

Thus the effect of epigenetic regulation is to change the probability of developmental pathways, while requiring more total energy for development. Hence the epigenetic information source Y acts as a *tunable catalyst*, a kind of second order cognitive enzyme, to enable and direct developmental pathways. This result permits hierarchical models similar to those of higher order cognitive neural function that incorporate contexts in a natural way (e.g., [98, 100]). The cost of this ability to channel is the metabolic necessity of supporting two information sources, X and Y, rather than just X itself.

This elaboration allows a spectrum of possible 'final' phenotypes, what S. Gilbert [43] calls developmental or phenotype plasticity. Thus gene expression is seen as, in part, responding to environmental or other, internal, developmental signals.

Including the effects of embedding culture in human ontology is, according to this formalism, straightforward: Consider culture as another embedding information source, Z, having source uncertainty $H(Z)$. Then the information chain rule becomes

$$H(X, Y, Z) \leq H(X) + H(Y) + H(Z) \tag{8}$$

and

$$\exp[nH(X, Y, Z)] \leq \exp[n(H(X) + H(Y) + H(Z))], \tag{9}$$

where, again, equality occurs only under stochastic independence.

A cultural regulatory apparatus, however, has very considerable free energy requirements, to grossly understate the matter.

In this model, following explicitly the direction indicated by Boyd, Kleinman, and their colleagues, culture is seen as an essential component of the catalytic epigenetic machinery that regulates human ontology, including development of the human mind. This is not to say that the development in other animals, particularly those that are highly social, does not undergo analogous regulation by larger scale structures of interaction. For human populations, however, social relations are themselves very highly regulated through an often strictly formalized cultural grammar and syntax.

5   The Groupoid Free Energy

A formal equivalence class algebra can now be constructed by choosing different origin and end points $\mathbf{S}_0, \mathbf{S}_\infty$ and defining equivalence of two states by the existence of a high probability meaningful path connecting them with the same origin and end. Disjoint partition by equivalence class, analogous to orbit equivalence classes for dynamical systems, defines the vertices of the proposed network of cognitive dual languages, much enlarged beyond the spinglass example. We thus envision a network of metanetworks. Each vertex then represents a different equivalence class of information sources dual to a cognitive process. This is an abstract set of metanetwork 'languages' dual to the cognitive processes of gene expression and development.

This structure generates a groupoid, in the sense of Weinstein [104]. States a_j, a_k in a set A are related by the groupoid morphism if and only if there exists a high probability grammatical path connecting them to the same base and end points, and tuning across the various possible ways in which that can happen – the different cognitive languages – parameterizes the set of equivalence relations and creates the (very large) groupoid. See the mathematical appendix for a summary of standard material on groupoids.

There is a hierarchy in groupoid structures. First, there is structure *within the system having the same base and end points*, as in [20, 21]. Second, there is a complicated groupoid structure defined by sets of dual information sources surrounding the variation of base and end points. We do not need to know what that structure is in any detail, but can show that its existence has profound implications.

First we examine the simple case, the set of dual information sources associated with a fixed pair of beginning and end states.

Taking the serial grammar/syntax model above, we find that not all high probability meaningful paths from $\mathbf{S}_0$ to $\mathbf{S}_\infty$ are the same. They are structured by the uncertainty of the associated dual information source, and that has a homological relation with free energy density.

Let us index possible dual information sources connecting base and end points by some set $A = \cup \alpha$. Argument by abduction from statistical physics is direct: Given metabolic energy density available at a rate M, and an allowed (fixed) characteristic development time τ, let $K = 1/\kappa M\tau$ for some appropriate scaling constant κ, so that $M\tau$ is total developmental free energy. Then we take the probability of a particular H_α as determined by a standard expression (e.g., [65]),

$$P[H_\beta] = \frac{\exp[-H_\beta K]}{\sum_\alpha \exp[-H_\alpha K]}, \tag{10}$$

where the sum may, in fact, be a complicated abstract integral.

This is just a version of the fundamental probability relation from statistical mechanics, as above. The sum in the denominator, the partition function in statistical physics, is a crucial normalizing factor that allows the definition of of $P[H_\beta]$ as a probability.

A basic requirement, then, is that the sum/integral always converges. K is the inverse product of a scaling factor, a metabolic energy density rate term, and a characteristic (presumed fixed) development time τ. The developmental energy might be raised to some power, e.g., $K = 1/(\kappa(M\tau)^b)$, suggesting the possibility of allometric scaling.

Some dual information sources will be 'richer'/smarter than others, but, conversely, will require more metabolic energy for their completion.

While we might simply impose an equivalence class structure based on equal levels of energy/source uncertainty, producing a groupoid, we can do more by now allowing both source and end points to vary, as well as by imposing energy-level equivalence. This produces a far more highly structured groupoid that we now investigate.

Equivalence classes define groupoids, by standard mechanisms [15, 47, 104]. The basic equivalence classes – here involving both information source uncertainty level and the variation of S_0 and S_∞, will define transitive groupoids, and higher order systems can be constructed by the union of transitive groupoids, having larger alphabets that allow more complicated statements in the sense of Ash above.

Again, given an appropriately scaled, dimensionless, fixed, inverse available metabolic energy density rate and development time, so that $K = 1/\kappa M\tau$, we propose that the metabolic-energy-constrained probability of an information source representing equivalence class D_i, H_{D_i}, will be given by the classic relation

$$P[H_{G_\alpha}] = \frac{\exp[-H_{G_\alpha}K]}{\sum_\beta \exp[-H_{G_\beta}K]},$$

where, now, we have shifted perspective, and *the sum/integral is over all possible elements of the largest available symmetry groupoid representing the equivalence class structure*. By the arguments of Ash above, compound sources, formed by the union of underlying transitive groupoids, being more complex, generally having richer alphabets, as it were, will all have higher free-energy-density-equivalents than those of the base (transitive) groupoids.

Let $Z_G \equiv \sum_\alpha \exp[-H_{G_\alpha}K]$. We now define the *Groupoid free energy* of the system, F_G, at inverse normalized metabolic energy density K, as

$$F_G[K] \equiv -\frac{1}{K} \log[Z_G[K]], \tag{11}$$

again following the standard arguments from statistical physics [38, 65].

5.1 Spontaneous Symmetry Breaking

The groupoid free energy permits introduction of an important idea from statistical physics.

We have expressed the probability of an information source in terms of its relation to a fixed, scaled, available (inverse) metabolic free energy density, seen as a kind of equivalent (inverse) system temperature. This gives a statistical thermodynamic path leading to definition of a 'higher' free energy construct – $F_G[K]$ – to which we now apply Landau's fundamental heuristic phase transition argument [65, 79, 89]. See, in particular, Pettini [79] for details.

Landau's insight was that second order phase transitions were usually in the context of a significant symmetry change in the physical states of a system, with one phase being far more symmetric than the other. A symmetry is lost in the transition, a phenomenon called spontaneous symmetry breaking, and symmetry changes are inherently punctuated. The greatest possible set of symmetries in a physical system is that of the Hamiltonian describing its energy states. Usually states accessible at lower temperatures will lack the symmetries available at higher temperatures, so that the lower temperature phase is less symmetric: The randomization of higher temperatures – in this case limited by available metabolic free energy densities – ensures that higher symmetry/energy states – mixed transitive groupoid structures – will then be accessible to the system. Absent high metabolic free energy rates and densities, however, only the simplest transitive groupoid structures can be manifest. A full treatment from this perspective seems to require invocation of groupoid representations, no small matter (e.g., [12, 16]).

Something like Pettini's [79] Morse-Theory-based topological hypothesis can now be invoked, i.e., that changes in underlying groupoid structure are a necessary (but not sufficient) consequence of phase changes in $F_G[K]$. Necessity, but not sufficiency, is important, as it, in theory, allows mixed groupoid symmetries.

Using this formulation, the mechanisms of epigenetic catalysis are accomplished by allowing the set B_1 above to span a distribution of possible 'final' states $\mathbf{S}_\infty$. Then the groupoid arguments merely expand to permit traverse of both initial states and possible final sets, recognizing that there can now be a possible overlap in the latter, and the epigenetic effects are realized through the joint uncertainties $H(X_{G_\alpha}, Z)$, so that the epigenetic information source Z serves to direct as well the possible final states of X_{G_α}. Again, Scherrer and Jost [86-7] use information theory arguments to suggest something similar.

5.2   The Groupoid Atlas

The groupoid free energy inherently defines a groupoid atlas in the sense of [7]. Following closely [45, 46], the set of groupoids G_α comprise a groupoid atlas $\mathcal{A}$ as follows.

A family of local groupoids $(G_\mathcal{A})$ is defined with respective object sets $(X_\mathcal{A})_\alpha$, and a *coordinate system* $\Phi_\mathcal{A}$ of $\mathcal{A}$ equipped with a reflexive relation $\leq$. These satisfy the following conditions:

(1). If $\alpha \leq \beta$ in $\Phi_\mathcal{A}$ then $(X_\mathcal{A})_\alpha \cap (X_\mathcal{A})_\beta$ is a union of components of $(G_\mathcal{A})$, that is, if $x \in (X_\mathcal{A})_\alpha \cap (X_\mathcal{A})_\beta$ and $g \in (G_\mathcal{A})_\alpha$ acts as $G : x \to y$, then $y \in (X_\mathcal{A})_\alpha \cap (X_\mathcal{A})_\beta$.

(2). If $\alpha \leq \beta$ in $\Phi_\mathcal{A}$, then there is a groupoid morphism defined between the restrictions of the local groupoids to intersections

$$(G_\mathcal{A})_\alpha | (X_\mathcal{A})_\alpha \cap (X_\mathcal{A})_\beta \to (G_\mathcal{A})_\beta | (X_\mathcal{A})_\alpha \cap (X_\mathcal{A})_\beta,$$

and which is the identity morphism on objects.

Thus each of the G_α with its associated dual information source H_{G_α} constitutes a component of an atlas that incorporates the dynamics of an interactive system by means of the intrinsic groupoid actions.

These are matters currently under very active study (e.g., [28]).

6 'Phase Change' and the Developmental Holonomy Groupoid in Phenotype Space

There is a more direct way to look at phase transitions in cognitive, and here culturally-driven, gene expression, adapting the topological perspectives of homotopy and holonomy directly within phenotype space.

We begin with ideas of directed homotopy.

In conventional topology one constructs equivalence classes of loops that can be continuously transformed into one another on a surface. The prospect of interest is to attempt to collapse such a family of loops to a point while remaining within the surface. If this cannot be done, there is a hole. Here we are concerned, as in figure 1, with sets of one-way developmental trajectories, beginning with an initial phenotype S_i, and converging on some final phenotype, here characteristic (highly dynamic) phenotypes labeled, respectively, S_n and S_o. One might view them as, respectively, 'normal' and 'other', and the developmental pathways as representing convergence on the two different configurations. The filled triangle represents the effect of a composite external epigenetic catalyst – including the effects of culture and culturally-structured social interaction – acting at a critical developmental period represented by the initial phenotype S_i.

We assume phenotype space to be directly measurable and to have a simple 'natural' metric defining the difference between developmental paths.

Developmental paths continuously transformable into each other without crossing the filled triangle define equivalence classes characteristic of different information sources dual to cognitive gene expression, as above.

Given a metric on phenotype space, and given equivalence classes of developmental trajectories having more than one path each, we can *pair one-way developmental trajectories* to make loop structures. In figure 1 the solid and

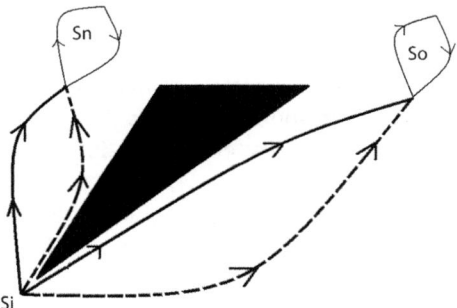

Fig. 1. Developmental homotopy equivalence classes in phenotype space. The set on one-way paths from S_i to S_n represents an equivalence class of developmental trajectories converging on a particular phenotype, here representing a highly dynamic normal structure. In the presence of a noxious external epigenetic catalyst, developmental trajectories can converge on a pathological structure, represented by the dynamic phenotype S_o.

dotted lines above and below the filled triangle can be pasted together to make loops characteristic of the different developmental equivalence classes. Although figure 1 is represented as topologically flat, there is no inherent reason for the phenotype manifold itself to be flat. The existence of a metric in phenotype space permits determining the degree of curvature, using standard methods. Figure 2 shows a loop in phenotype space. Using the metric definition it is possible to *parallel transport* a tangent vector starting at point s around the loop, and to measure the angle between the initial and final vectors, as indicated. A central result from elementary metric geometry is that the angle α will be given by the integral of the curvature tensor of the metric over the interior of the loop (e.g., [40], Section 9.6).

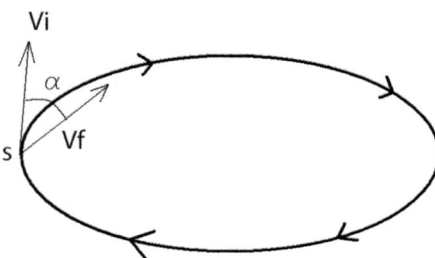

Fig. 2. Parallel transport of a tangent vector $V_i \rightarrow V_f$ around a loop on a manifold. Only for a geometrically flat object will the angle between the initial and final vectors be zero. By a fundamental theorem the path integral around the loop by parallel displacement is the surface integral of the curvature over the loop.

The *holonomy group* is defined as follows (e.g., [50]):

If s is a point on a manifold M having a metric, then the holonomy group of M is the group of all linear transformations of the tangent space M_s obtained by parallel translation along closed curves starting at s.

For figure 1 the *phenotype holonomy groupoid* is the disjoint union of the different holonomy groups corresponding to the different branches separated by 'developmental shadows' induced by epigenetic information sources acting as developmental catalysts.

The relation between the phenotype groupoid as defined here and the phase transitions in $F_G[K]$ as defined above is an open question, and is a central focus of ongoing work.

7   Holonomy on the Manifold of Dual Information Sources

7.1   Basic Structure

Glazebrook and Wallace [45] examined holonomy groupoid phase transition arguments for networks of interacting information sources dual to cognitive phenomena. A more elementary form of this arises directly through extending holonomy groupoid arguments to a manifold of different information sources dual to cognitive phenomena as follows.

Different cognitive phenomena will have different dual information sources, and we are interested in the local properties of the system near a particular reference state. We impose a topology on the system, so that, near a particular 'language' A, dual to an underlying cognitive process, there is an open set U of closely similar languages $\hat{A}$, such that $A, \hat{A} \subset U$. It may be necessary to coarse-grain the system's responses to define these information sources. The problem is to proceed in such a way as to preserve the underlying essential topology, while eliminating 'high frequency noise'. The formal tools for this can be found elsewhere, e.g., in Chapter 8 of [17].

Since the information sources dual to the cognitive processes are similar, for all pairs of languages $A, \hat{A}$ in U, it is possible to:

(1). Create an embedding alphabet which includes all symbols allowed to both of them.

(2). Define an information-theoretic distortion measure in that extended, joint alphabet between any high probability (grammatical and syntactical) paths in A and $\hat{A}$, which we write as $d(Ax, \hat{A}x)$ [25]. More detail on distortion measures is given in the section below on the Rate Distortion Theorem. Note that these languages do not interact, in this approximation.

(3). Define a metric on U, for example,

$$\mathcal{M}(A, \hat{A}) = |\lim \frac{\int_{A,\hat{A}} d(Ax, \hat{A}x)}{\int_{A,A} d(Ax, A\hat{x})} - 1|, \tag{12}$$

integrating over the sets of high probability paths. Note that the integration in the denominator is over different paths within A itself, while in the numerator it is between different paths in A and $\hat{A}$. Other metric constructions on U seem possible, leading to similar results, just as different definitions of distortion lead to the same end in the Rate Distortion Theorem.

Structures weaker than a conventional metric would be of more general utility, but the mathematical complications are formidable.

Note that these conditions can be used to define equivalence classes of *languages* dual to cognitive processes, where previously we defined equivalence classes of *states* that could be linked by high probability, grammatical and syntactical paths connecting two phenotypes. This led to the characterization of different information sources. Here we construct an entity, formally a topological manifold, *that is an equivalence class of information sources*. This is, provided $\mathcal{M}$ is a conventional metric, a classic differentiable manifold. The set of such equivalence classes generates the *dynamical groupoid*, and questions arise regarding mechanisms, internal or external, which can break that groupoid symmetry.

Since H and $\mathcal{M}$ are both scalars, a 'covariant' derivative can be defined directly as

$$dH/d\mathcal{M} = \lim_{\hat{A} \to A} \frac{H(A) - H(\hat{A})}{\mathcal{M}(A, \hat{A})}, \tag{13}$$

where $H(A)$ is the source uncertainty of language A.

The essential point of a 'covariant' derivative is that it is *independent of coordinate systems*, a condition this definition fulfills. As shown above, this leads directly to ideas of a derivative along a tangent vector and to ideas of parallel transport leading to deep topological concepts such as holonomy. Introduction of a coordinate system in the definition of $\mathcal{M}$ quickly leads to the usual Christoffel symbols and the familiar geodesic equations (e.g., [98], Section 8.3).

Suppose the system to be set in some reference configuration A_0.

To obtain the unperturbed dynamics of that state, impose a Legendre transform using this derivative, defining another scalar

$$S \equiv H - \mathcal{M}dH/d\mathcal{M}. \tag{14}$$

The simplest possible generalized Onsager relation – here seen as an empirical, fitted, equation like a regression model – is

$$d\mathcal{M}/dt = LdS/d\mathcal{M}, \tag{15}$$

where t is the time and $dS/d\mathcal{M}$ represents an analog to the thermodynamic force in a chemical system. This is seen as acting on the reference state A_0.

Again, explicit parameterization of $\mathcal{M}$ – that is, imposing a coordinate system – introduces standard, and quite considerable, notational complications [17]. Defining a metric for different cognitive dual languages parameterized by $\mathbf{K}$ leads to Riemannian, or even Finsler, geometries, including the usual geodesics [45, 46, 98].

The dynamics, as we have presented them so far, have been noiseless. The simplest generalized Onsager relation in the presence of noise might be rewritten as

$$d\mathcal{M}/dt = LdS/d\mathcal{M} + \sigma W(t),$$

where σ is a constant and $W(t)$ represents white noise. Again, S is seen as a function of the parameter $\mathcal{M}$. This leads directly to a family of classic stochastic differential equations of the form

$$d\mathcal{M}_t = L(t, \mathcal{M})dt + \sigma(t, \mathcal{M})dB_t, \tag{16}$$

where L and σ are appropriately regular functions of t and $\mathcal{M}$, and dB_t represents the noise structure, characterized by its quadratic variation. In the sense of Emery [37], this leads into complicated realms of stochastic differential geometry and related topics.

The natural generalization is to a system of developmental processes that influence each other via mutual information crosstalk, as described by [101].

7.2 'Coevolutionary' Development

Here we examine multiple interacting information sources representing simultaneous gene expression processes. This is, in a broad sense, a 'coevolutionary' phenomenon in that the development of one process may affect that of others.

Most generally we require that different cognitive developmental subprocesses of gene expression characterized by information sources H_m interact through chemical or other signals and assume that *different processes become each other's principal environments*.

We write

$$H_m = H_m(K_1...K_s, ...H_j...), \tag{17}$$

where the K_s represent other relevant parameters and $j \neq m$.

The dynamics of such a system is driven by a recursive network of stochastic differential equations, similar to those used to study many other highly parallel dynamic structures (e.g., [107]).

Letting the K_j and H_m all be represented as parameters Q_j, (with the caveat that H_m not depend on itself), one can define, according to the generalized Onsager development of [100, 101],

$$S^m = H_m - \sum_i Q_i \partial H_m / \partial Q_i$$

to obtain a complicated recursive system of phenomenological 'Onsager relations' stochastic differential equations,

$$dQ_t^j = \sum_i [L_{j,i}(t, ...\partial S^m / \partial Q^i...)dt + \sigma_{j,i}(t, ...\partial S^m / \partial Q^i...)dB_t^i], \tag{18}$$

where, again for notational simplicity only, we have expressed both the H_j and the external K's in terms of the same symbols Q_j.

m ranges over the H_m and we could allow different kinds of 'noise' dB_t^i, having particular forms of quadratic variation that may, in fact, represent a projection of environmental factors under something like a rate distortion manifold [46].

It is important to realize that, for this formulation, one does not necessarily have the equivalent of 'Onsager's fourth law' of thermodynamics, i.e., the symmetry relation $L_{i,j} = L_{j,i}$. This is because such a symmetry, at base, is a statement of local time reversal invariance (e.g., [26], pp. 35-41). But information sources are notoriously one-way in time, for example someone speaking or writing in English is much more likely to utter the five-character string " the " than its reverse. More generally, information sources are characterized by their 'order', the number of sequential steps over which serial correlations can be observed. Rich information sources, representing complicated cognitive phenomena, can be of very high order indeed. This suggests a 'weaker' structure for empirical Onsager relations observed for cognitive processes than would be expected for relatively simple physical phenomena.

As usual, for a system of equations like (18), there will be multiple quasi-stable points, representing a class of generalized resilience modes accessible via holonomy punctuation.

There are, indeed, many possible patterns:

(1). Setting equation (18) equal to zero and solving for stationary points gives attractor states since the noise terms preclude unstable equilibria.

(2). This system may, however, converge to limit cycle or 'strange attractors' that are very highly dynamic.

(3). What is converged to in both cases is not a simple state or limit cycle of states. Rather it is an equivalence class, or set of them, of generalized language information sources coupled by mutual interaction through crosstalk. Thus 'stability' in this extended model represents particular patterns of ongoing dynamics rather than some identifiable 'state', although such dynamics may be indexed by a 'stable' set of phenotypes.

Here we become enmeshed in a system of highly recursive phenomenological stochastic differential equations, but at a deeper level than the standard stochastic chemical reaction model (e.g., [107]), and in a dynamic rather than static manner: the objects of this system are equivalence classes of information sources and their crosstalk, rather than simple final states of a chemical system.

We have defined a groupoid for the system based on a particular set of equivalence classes of information sources dual to cognitive processes. That groupoid parsimoniously characterizes the available dynamical manifolds, and breaking of the groupoid symmetry by epigenetic crosstalk creates more complex objects of considerable interest. This leads to the possibility, indeed, the necessity of epigenetic *Deus ex Machina* mechanisms – analogous to programming, stochastic resonance, etc. – to force transitions between the different possible modes within and across dynamic manifolds. In one model the external 'programmer' creates the manifold structure, and the system hunts within that structure for the 'solution' to the problem according to equivalence classes of paths on the manifold. Noise, as with random mutation in evolutionary algorithms, precludes simple unstable equilibria, but not other possible structures.

Equivalence classes of *states* gave dual information sources. Equivalence classes of *information sources* give different characteristic dynamical manifolds. Equivalence classes of one-way developmental *paths* produce different directed homotopy topologies characterizing those dynamical manifolds. This introduces the possibility of having different quasi-stable modes *within* individual manifolds, and leads to ideas of holonomy and the holonomy groupoid of the set of quasi-stable developmental modes.

8   Toward Empirical Tests of Theory: The Rate Distortion Models

We have introduced a spectrum of abstract models of development and its pathologies founded on a cognitive paradigm for gene expression that is itself based on application of the asymptotic limit theorems of information theory. Following the classic pattern of parametric statistics arising via the Central Limit Theorem, we would like to see tools emerge that would allow both the analysis of empirical or observational data and the robust testing of theory. The simplest first step for such a program seems to lie in the the direction of the generalized Onsager relations that characterize system dynamics. Here we present a restriction of the theory that may prove useful for empirical comparisons.

8.1   The Rate Distortion Theorem

The interaction between cognitive structures can be restated from a highly simplified perspective via the Rate Distortion theorem. Suppose a sequence of signals is generated by an information source dual to a cognitive process, Y having output $y^n = y_1, y_2, ...$ This is 'digitized' in terms of the observed behavior of the system with which it communicates, say a sequence of observed behaviors $b^n = b_1, b_2, ...$ Often the b_i will happen in a characteristic 'real time' τ. Assume each b^n is then deterministically retranslated back into a reproduction of the original biological signal,

$$b^n \rightarrow \hat{y}^n = \hat{y}_1, \hat{y}_2, ...$$

Define a distortion measure $d(y, \hat{y})$ that compares the original to the retranslated path. Many such measures are possible. The Hamming distortion, for example, is

$$d(y, \hat{y}) = 1, y \neq \hat{y}$$

$$d(y, \hat{y}) = 0, y = \hat{y}$$

For continuous variates the squared error distortion is

$$d(y, \hat{y}) = (y - \hat{y})^2.$$

The distortion between *paths* y^n and $\hat{y}^n$ is defined as

$$d(y^n, \hat{y}^n) \equiv \frac{1}{n} \sum_{j=1}^{n} d(y_j, \hat{y}_j).$$

A remarkable fact of the Rate Distortion Theorem is that *the basic result is independent of the exact distortion measure chosen* [25, 29].

Suppose that with each path y^n and b^n-path retranslation into the y-language, denoted $\hat{y}^n$, there are associated individual, joint, and conditional probability distributions

$$p(y^n), p(\hat{y}^n), p(y^n, \hat{y}^n), p(y^n | \hat{y}^n).$$

The average distortion is defined as

$$D \equiv \sum_{y^n} p(y^n) d(y^n, \hat{y}^n). \tag{19}$$

It is possible, using the distributions given above, to define the information transmitted from the Y to the $\hat{Y}$ process using the Shannon source uncertainty of the strings:

$$I(Y,\hat{Y}) \equiv H(Y) - H(Y|\hat{Y}) = H(Y) + H(\hat{Y}) - H(Y,\hat{Y}), \qquad (20)$$

where $H(...,...)$ is the joint and $H(...|...)$ the conditional uncertainty [2, 25].

If there is no uncertainty in Y given the retranslation $\hat{Y}$, then no information is lost, and the systems are in perfect synchrony.

In general, of course, this will not be true.

The *rate distortion function* $R(D)$ for a source Y with a distortion measure $d(y,\hat{y})$ is defined as

$$R(D) = \min_{p(y,\hat{y}); \sum_{(y,\hat{y})} p(y)p(y|\hat{y})d(y,\hat{y}) \leq D} I(Y,\hat{Y}). \qquad (21)$$

The minimization is over all conditional distributions $p(y|\hat{y})$ for which the joint distribution $p(y,\hat{y}) = p(y)p(y|\hat{y})$ satisfies the average distortion constraint (i.e., average distortion $\leq D$).

The *Rate Distortion Theorem* states that $R(D)$ is the minimum necessary rate of information transmission which ensures the communication between the modules does not exceed average distortion D. Thus $R(D)$ defines a minimum necessary channel capacity. Cover and Thomas [25] or Dembo and Zeitouni [29] provide details. The rate distortion function has been calculated for a number of systems.

There is an absolutely central fact characterizing the rate distortion function: Cover and Thomas ([25], Lemma 13.4.1) show that $R(D)$ *is necessarily a decreasing convex function of* D for any reasonable definition of distortion.

That is, $R(D)$ *is always* a reverse J-shaped curve. This will prove crucial for the overall argument. Indeed, convexity is an exceedingly powerful mathematical condition, and permits deep inference (e.g., [83]). Ellis ([36], Ch. VI) applies convexity theory to conventional statistical mechanics.

For a Gaussian channel having noise with zero mean and variance σ^2 [25],

$$R(D) = 1/2 \log[\sigma^2/D], 0 \leq D \leq \sigma^2,$$
$$R(D) = 0, D > \sigma^2. \qquad (22)$$

Recall, now, the relation between information source uncertainty and channel capacity (e.g., [2]):

$$H[\mathbf{X}] \leq C, \qquad (23)$$

where H is the uncertainty of the source X and C the channel capacity, defined according to the relation [2]

$$C \equiv \max_{P(X)} I(X|Y), \qquad (24)$$

where $P(X)$ is chosen so as to maximize the rate of information transmission along a channel Y.

Finally, recall the analogous definition of the rate distortion function from equation (21), again an extremum over a probability distribution.

8.2   Rate Distortion Dynamics

$R(D)$ defines the minimum channel capacity necessary for the system to have average distortion less than or equal D, placing a limit on information source uncertainty. Thus, we suggest distortion measures can drive information system dynamics. That is, the rate distortion function also has a homological relation to free energy density, similar to the relation between free energy density and information source uncertainty.

We are led to propose, as a heuristic, that the dynamics of cognitive modules interacting in a characteristic 'real time' τ will be constrained by the system as described in terms of a parameterized rate distortion function. To do this, take R as parameterized, not only by the distortion D, but by some vector of variates $\mathbf{Q} = (Q_1, ..., Q_k)$, for which the first component is the average distortion. The assumed dynamics are, as in [100, 101], then driven by gradients in the rate distortion disorder defined as

$$S_R \equiv R(\mathbf{Q}) - \sum_{i=1}^{k} Q_i \partial R / \partial Q_i. \qquad (25)$$

This leads to the deterministic and stochastic systems of equations analogous to the Onsager relations of nonequilibrium thermodynamics:

$$dQ_j/dt = \sum_i L_{j,i} \partial S_R / \partial Q_i \qquad (26)$$

and

$$dQ_t^j = L^j(Q_1, ..., Q_k, t)dt + \sum_i \sigma^{j,i}(Q_1, ..., Q_k, t)dB_t^i, \qquad (27)$$

where the dB_t^i represent added, often highly structured, stochastic 'noise' whose properties are characterized by the quadratic variation (e.g., [81]).

Even for this simplified structure, it is not clear under what circumstances 'Onsager reciprocal relations' are possible. Since average distortion is a scalar, however, some systems may indeed display the kind of time reversal invariance required for those symmetries.

A central focus of this paper, however, is to generalize these equations in the face of richer structures, for example interactions between cognitive modules that may not be time-reversible, the existence of characteristic time constants within nested processes, and the influence of an embedding source of free energy.

For a simple Gaussian channel with noise having zero mean and variance σ^2,

$$S_R(D) = R(D) - DdR(D)/dD = 1/2 \log(\sigma^2/D) + 1/2. \qquad (28)$$

The simplest possible Onsager relation becomes

$$dD/dt = -\mu dS_R/dD = \frac{\mu}{2D}, \tag{29}$$

where $-dS_R/dD$ represents the force of an entropic wind, a kind of internal dissipation inevitably driving the real-time, system of interacting (cognitive) information sources toward greater distortion.

This has the solution

$$D = \sqrt{\mu t}, \tag{30}$$

so that the average distortion increases monotonically with time, for this model.

A central observation is that *similar results must necessarily apply to any of the reverse-J-shaped relations that inevitably characterize $R(D)$*, since the rate distortion function is necessarily a convex decreasing function of the average distortion D, whatever distortion measure is chosen. Again, see [25], (Lemma 13.4.1) for details.

The explicit implication is that a system of cognitive modules interacting in real time will inevitably be subject to a relentless entropic force, requiring a constant free energy expenditure for maintenance of some fixed average distortion in the communication between them: The distortion in the communication between two interacting modules will, without free energy input, have time dependence

$$D = f(t), \tag{31}$$

with $f(t)$ monotonic increasing in t.

This necessarily leads to the punctuated failure of the system.

Note that equation (30) is similar to classical Brownian motion as treated by Einstein: Let $p(x, t)dx$ be the probability a particle located at the origin at time zero and undergoing Brownian motion is found at locations $x \to x + dx$ at time t. Then, p satisfies the diffusion equation

$$\partial p(x, t)/\partial t = \mu \partial^2 p(x, t)/\partial x^2.$$

Einstein's solution is that

$$p(x, t) = \frac{1}{\sqrt{4\pi \mu t}} \exp[-x^2/4\mu t].$$

It is easy to show that the standard deviation of the particle position increases in proportion to $\sqrt{\mu t}$, just as above.

Some comment is appropriate. Following, e.g., [19], a process $B = B_t, t \in \mathcal{R}_+$ is called a Brownian motion in $\mathcal{R}_+$ iff:

(1). for $0 \leq s < t < \infty$, $B_t - B_s$ is a normally distributed random variate with mean zero and variance $|t - s|$.

(2). for $0 \leq t_0 < t_1 < ... < t_k < \infty$,

$$\{B_{t_0}; B_{t_j} - B_{t_{j-1}}, j = 1, ..., k\}$$

is a set of independent random variates.

An information source, of course, generates a *highly correlated sequence* that grossly violates these simple assumptions. What we have shown is that the *distortion* in the communication between interacting cognitive modules, under appropriate empirical Onsager relations, can behave as if it were undergoing Brownian motion.

This is a simple, but far from trivial, result.

Prandolini and Moody [80] have, in fact, observed something much like this in the time base error of recorded signals. Wow and flutter are the instantaneous speed error between recording and reproduction epochs. The time base error (TBE) in the reproduced signal is a function of the wow and flutter. They show, empirically, that the nonperiodic TBE is a *fractional Brownian motion*. The implication is that the nonperiodic flutter is fractional Gaussian, and thus what they call a 'blind' TBE system is impractical for the design of a TBE compensation system.

Normalized fractional Brownian motion on $(0, t), t \in \mathcal{R}_+$ is a continuous time Gaussian process starting at zero, with mean zero, and having the covariance function [11]

$$E[B^H(t)B^H(s)] = (1/2)[|t|^{2H} + |s|^{2H} - |t - s|^{2H}].$$

If $H = 1/2$ the process is a regular Brownian motion. Otherwise, for $H > 1/2$, the increments are positively correlated, and for $H < 1/2$, negatively correlated.

We will explore this kind of relation in more detail below.

8.3   Rate Distortion Coevolutionary Dynamics

A simplified version of equation (18) can be constructed using the rate distortion functions for mutual crosstalk between a set of interacting cognitive modules, using the homology of the rate distortion function itself with free energy, as driven by the inherent convexity of the Rate Distortion Function $R(D)$. That convexity is, in fact, why we invoke the Rate Distortion Function.

Given different cognitive processes $1...s$, the quantities of special interest thus become the mutual rate distortion functions $R_{i,j}$ characterizing communication (and the distortion $D_{i,j}$) between them, while the essential parameters remain the characteristic time constants of each process, $\tau_j, j = 1...s$, and an overall, embedding, available free energy density, F.

Taking the Q^α to run over all the relevant parameters and mutual rate distortion functions (including distortion measures $D_{i,j}$), equation (14) becomes

$$S_R^{i,j} \equiv R_{i,j} - \sum_k Q^k \partial R_{i,j}/\partial Q^k. \tag{32}$$

Equation (18) accordingly becomes

$$dQ_t^\alpha = \sum_{\beta=(i,j)} [L_\beta(t, ...\partial S_R^\beta/\partial Q^\alpha...)dt + \sigma_\beta(t, ...\partial S_R^\beta/\partial Q^\alpha...)dB_t^\beta], \tag{33}$$

and this generalizes the treatment in terms of crosstalk, its distortion, the inherent time constants of the different cognitive modules, and the overall available free energy density.

This is a very complicated structure indeed, but its general dynamical behaviors will obviously be analogous to those described just above. For example, setting equation (33) to zero gives the 'coevolutionary stable states' of a system of interacting cognitive modules. Again, limit cycles and strange attractors seem possible as well. And again, what is converged to is a dynamic behavior pattern, not some fixed 'state'. And again, such a system will display highly punctuated dynamics almost exactly akin to resilience domain shifts in ecosystems [48, 52]. Indeed, the formalism seems directly applicable to ecosystem studies.

And again, because these are highly self-dynamic cognitive phenomena and not simple crystals or other physical objects, it may not often be possible to invoke time reversal invariance to give Onsager-like reciprocal symmetries.

8.4   Some Examples

The Gaussian channel. First, assume a fixed embedding communication free energy density of F, representing the richness of incoming information from the interacting cognitive modules. The simplest generalization of equation (29), for a Gaussian channel, becomes

$$dD/dt = \mu/2D - \alpha F, \alpha > 0, \tag{34}$$

characterizing the distortion dynamics.

This has the equilibrium solution

$$D_{equlib} = \frac{\mu}{2\alpha F}. \tag{35}$$

In contrast to equation (29), where, in the absence of some free energy/ information input, the distortion grows as the square root of the elapsed time, here there is a finite, equilibrium, average distortion, that is inversely proportional to the available environmental or informational free energy, that the interacting systems can use to direct their actions.

The obvious generalization is

$$D_{equilib} = \frac{1}{g(F)}, \tag{36}$$

where $g(F)$ is monotonic increasing in F.

Introducing a characteristic response time variable τ, so that

$$dD/dt = \frac{\mu}{2D} - g(F)h(\tau), \tag{37}$$

where $h(\tau)$ is also monotonic increasing, leads to

$$D_{equilib} = \frac{\mu}{2g(F)h(\tau)}. \tag{38}$$

Thus, for this particular phenomenological Onsager model, at a fixed rate of available information free energy, increasing allowable response time decreases average distortion in the interaction between the cognitive subsystems.

This is, in fact, a classic result across a broad spectrum of engineering applications.

If we now allow feedback, so that the system actively seeks information in proportion to the distortion between intent and impact, then the empirical Onsager relation for a Gaussian channel becomes

$$dD/dt = \frac{\mu}{2D} - g(F)h(\tau)D, \tag{39}$$

and

$$D_{equilib} = \sqrt{\frac{\mu}{2g(F)h(\tau)}}, \tag{40}$$

significantly smaller than (38).

This is, in fact, precisely the classic result for Brownian motion in a harmonic central field (e.g., [103], eq. 54), restated in terms of average distortion.

A mixed strategy,

$$dD/dt = \frac{\mu}{2D} - g(F)h(\tau)[1 + \alpha D], \tag{41}$$

leading to the quadratic

$$2Dg(F)h(\tau)(1 + \alpha D) - \mu = 0, \tag{42}$$

has a single equilibrium solution

$$D_{equilib} = \frac{-g(F)h(\tau) + \sqrt{g(F)^2 h(\tau)^2 + 2g(F)h(\tau)\alpha\mu}}{2g(F)h(\tau)\alpha}, \tag{43}$$

since D must be greater than zero and real.

Stochastic generalizations – the diffusion of distortion as it were – involving probabilistic fuzz about deterministic equilibria or dynamic paths, seem direct.

The 'Natural' channel. According to [84], operational rate-distortion functions of most natural images, when compressed with state-of-the-art wavelet coders, exhibit power-law behavior, rather than the logarithmic function of a Gaussian channel, so that

$$R(D) = \frac{\beta}{D^\gamma}, \tag{44}$$

usually with $\gamma \approx 1$.

Applying our formalism to such a 'natural' channel gives

$$S(D) = R(D) - DdR/dD = \frac{\beta(1 + \gamma)}{D^\gamma},$$

$$dD/dt = -\mu dS/dD = \frac{\mu\beta\gamma(1+\gamma)}{D^{1+\gamma}},$$

having the solution

$$D(t) = [t\mu\beta(2+\gamma)(1+\gamma)]^{1/(2+\gamma)},$$

which, for $\gamma \approx 1$, becomes

$$D(t) \approx (6\beta\mu t)^{1/3}.$$

Thus distortion would grow approximately as the cube root of time for such a system if it were to undergo a self-dynamic process.

Taking $\gamma = 1$ and introducing free energy and characteristic time terms $g(F)$ and $h(\tau)$ as above

leads to dynamic equations

$$dD/dt = \frac{2\mu\beta}{D^2} - g(F)h(\tau)$$
$$dD/dt = \frac{2\mu\beta}{D^2} - g(F)h(\tau)D,$$
(45)

having equilibrium solutions

$$D_{equilib} = \sqrt{\frac{2\mu\beta}{g(F)h(\tau)}}$$
$$D_{equilib} = [\frac{2\mu\beta}{g(F)h(\tau)}]^{1/3}.$$
(46)

Distributed information input. A somewhat different picture emerges if we permit input from a one parameter *distribution* of energy/information sources. Defining $\hat{F} = g(F)h(\tau)$, we have for the Gaussian and Natural channels respectively, with $R(D) = 1/2\log(\sigma^2/D), \beta/D^\gamma, \gamma \approx 1$, the empirical Onsager relations

$$dD/dt = \frac{\mu}{2D} - <\hat{F}>, \frac{\mu}{2D} - <\hat{F}> D$$
(47)

and

$$dD/dt = \frac{2\mu\beta}{D^2} - <\hat{F}>, \frac{2\mu\beta}{D^2} - <\hat{F}> D,$$
(48)

where $<\hat{F}>$ is the average over the distribution of incoming information/free energy. The inherently convex nature of $R(D)$ ensures roughly similar relations in general.

Typically we can assign some 'effective temperature', say T, to the incoming distribution so that

$$<\hat{F}> = \frac{\int_0^\infty \hat{F}\exp[-\hat{F}/kT]d\hat{F}}{\int_0^\infty \exp[-\hat{F}/kT]d\hat{F}} = kT.$$
(49)

This leads to equilibrium distortions

$$D_{equilib} = \frac{\mu}{2kT}, \sqrt{\frac{\mu}{2kT}} \tag{50}$$

for the Gaussian channel, and

$$D_{equilib} = \sqrt{\frac{2\mu\beta}{kT}}, [\frac{2\mu\beta}{kT}]^{1/3} \tag{51}$$

for the Natural channel.

Given a finite maximum for $g(F)h(\tau) \equiv M$, then

$$< \hat{F} >= \frac{kT[\exp(M/kT) - 1] - M}{\exp(M/kT) - 1}, \tag{52}$$

and the expressions for equilibrium distortion must be adjusted accordingly.

Typically, a cognitive system will have a distribution of information inputs having varying fidelity/energy, and some sort of averaging across them would be expected. The 'effective temperature' approach seems most direct.

These calculations suggest that a simple empirical Onsager treatment may be quite powerful.

9   Expanding the Mathematical Approach

We have, in the context of the tunable epigenetic catalysis of [101], developed three separate phase transition/branching models of cognitive gene expression based on groupoid structures that may be applied to development and its dysfunctions, as known to be particularly influenced by embedding culture. The first used Landau's spontaneous symmetry breaking to explore phase transitions in a groupoid free energy $F_G[K]$. The second examined a holonomy groupoid in phenotype space generated by disjoint developmental homotopy equivalence classes, and 'loops' constructed by pairing one-way development paths. The third introduced a metric on a manifold of different information sources dual to cognitive gene expression, leading to a more conventional picture of parallel transport around a loop leading to holonomy. The dynamical groupoid of [98] (Sec. 3.8) is seen as involving a disjoint union across underlying manifolds that produces a holonomy groupoid in a natural manner.

There are a number of outstanding mathematical questions.

The first is the relation between the Landau formalism and the structures of phenotype space S and those of the associated manifold of dual information sources, the manifold M having metric $\mathcal{M}$. How does epigenetic catalysis in M-space imposes structure on S-space? How is this related to spontaneous symmetry breaking?

What would a stochastic version of the theory, in the sense of [37], look like? It is quite possible, using appropriate averages of the stochastic differential equations that arise naturally, to define parallel transport, holonomy, and the like for these structures. In particular a stochastic extension of the results of the first question would seem both fairly direct and interesting from a real-world perspective, as development is always 'noisy'.

The construction of loops from directed homotopy arcs in figure 1 is complicated by the necessity of imposing a consistent piecewise patching rule for parallel translation at the end of each arc, say from $\mathbf{S}_i$ to $\mathbf{S}_n$. This can probably be done by some standard fiat, but the details will likely be messy.

On another matter, we have imposed metrics on S and M space, making possible a fairly standard manifold analysis of complex cognitive processes of gene expression and development. While this is no small thing, the 'natural' generalization, given the ubiquity of groupoids across our formalism, would be to a more complete groupoid atlas treatment in the spirit of Section 5.2. The groupoid atlas permits a weaker structure compared with that of a conventional manifold since no condition of compatibility between arbitrary overlaps of the patches is necessary. It is possible that the groupoid atlas will become, to complicated problems in biological cognitive process, something of what the Riemannian manifold has been to physics.

With regard to questions of 'smoothness', we are assuming that the cognitive landscape of gene expression is sufficiently rich that discrete paths can be well approximated as continuous where necessary, the usual physicist's hack.

Finally, sections 6 and 7 are based on existence of more-or-less conventional metrics, and this may not be a good approximation to many real systems. Extending topological phase transition theory to 'weaker' topologies, e.g., Finsler geometries and the like, is not a trivial task.

10   Discussion

We began with an exploration of the role of culture in mental disorders – quintessential developmental dysfunctions – and, inferring the general from the particular, have expanded the perspective to a spectrum of broadly heritable diseases. Culturally structured psychosocial stress, and similar noxious exposures, can write distorted images of themselves onto human ontology – both child growth, and, if sufficiently powerful, adult development as well – by a variety of mechanisms, initiating a punctuated trajectory to characteristic forms of comorbid mind/body dysfunction. This occurs in a manner recognizably analogous to resilience domain shifts affecting stressed ecosystems [48, 52, 97]. Consequently, like ecosystem restoration, reversal or palliation may often be exceedingly difficult once a generalized domain shift has taken place. Thus a public health approach may be paramount: rather than seeking to understand why half a population does not respond to the LD_{50} of a teratogenic environmental exposure, one seeks policies and social reforms that limit the exposure.

Both socio-cultural and epigenetic environmental influences – like gene methylation – are heritable, in addition to genetic mechanisms. The missing heritability of complex diseases that Manolio et al. [66] seek to find in more and better gene studies is most likely dispersed within the 'dark matter' of these two other systems of heritage that together constitute the larger, and likely highly synergistic, regulatory machinery for gene expression. More and more purely genetic studies would, under such circumstances, be akin to using increasingly powerful microscopes to look for cosmic membranes of strewn galaxies.

A crucial matter is the conversion of the probability models we present here into statistical tools suitable for analyzing real data, and hence actually testing the theoretical structure. Some work in this direction has been done in Section 8, but the problem involves not just programming such models for use, but identifying appropriate real-world situations, assembling available data sets, transforming the data as needed for the models, and actually applying them and comparing the model predictions to data. Indeed, the environmental health literature contains numerous examples of developmental deviations due to either chemical exposures or interaction between chemical and socioeconomic exposures, and these could serve as sources of data for direct analysis (e.g., [9, 22, 30, 41, 44, 58, 61, 73, 74, 85, 88, 93, 94, 99]). Thus, quite a number of data sets exist in the environmental health and socioeconomic epidemiological literature that could be subjected to meta-analysis and other review for model verification and fitting. Our topological models, when converted to statistical tools for data analysis, hold great potential for understanding developmental trajectories and interfering factors (teratogens) through the life course. Sets of cross cultural variants of these data would be needed to address the particular concerns of this paper. In addition, as the previous section indicates, unsolved mathematical questions abound.

Nonetheless, what we have done is of no small interest for understanding human ontology, its pathologies, and their heritability. West-Eberhard [105, 106] argues that any new input, whether it comes from the genome, like a mutation, or from the external environment, like a temperature change, a pathogen, or a parental opinion, has a developmental effect only if the preexisting phenotype is responsive to it. A new input causes a reorganization of the phenotype, or 'developmental recombination'. In developmental recombination, phenotypic traits are expressed in new or distinctive combinations during ontogeny, or undergo correlated quantitative change in dimensions. Developmental recombination can result in evolutionary divergence at all levels of organization.

According to West-Eberhard, individual development can be visualized as a series of branching pathways. Each branch point is a developmental decision, or switch point, governed by some regulatory apparatus, and each switch point defines a modular trait. Developmental recombination implies the origin or deletion of a branch and a new or lost modular trait. The novel regulatory response and the novel trait originate simultaneously. Their origins are, in fact, inseparable events: There cannot, West-Eberhard concludes, be a change in the phenotype, a novel phenotypic state, without an altered developmental pathway.

Our analysis provides a new formal picture of this process as it applies to human development: The normal branching of developmental trajectories, and the disruptive impacts of teratogeneic events of various kinds, can be described in terms of a growing sequence of holonomy groupoids, each associated with a set of dual information sources representing patterns of cognitive gene expression catalyzed by epigenetic information sources that, for humans, must include culture and culturally-modulated social interaction as well as more direct mechanisms like gene methylation. This is a novel way of looking at human development and its disorders that may prove to be of some use. The most important innovation of this work, however, seems to be the natural incorporation of embedding culture as an essential component of the epigenetic regulation of human ontology, and in the effects of environment on the expression of a broad spectrum of developmental disorders: the missing heritability of complex diseases found.

Acknowledgments

The author thanks M. Weissman for detailed criticisms that helped sharpen the argument, S. Heine for access to several preprints, and two anonymous reviewers for comments useful in revision.

References

1. Arnett, J.: The neglected 95 %. The American Psychologist 63, 602–614 (2008)
2. Ash, R.: Information Theory. Dover Publications, New York (1990)
3. Atlan, H., Cohen, I.: Immune information, self-organization, and meaning. International Immunology 10, 711–717 (1998)
4. Atmanspacher, H.: Toward an information theoretical implementation of contextual conditions for consciousness. Acta Biotheoretica 54, 157–160 (2006)
5. Baars, B.: A Cognitive Theory of Consciousness. Cambridge University Press, New York (1989)
6. Backdahl, L., Bushell, A., Beck, S.: Inflammatory signalling as mediator of epigenetic modulation in tissue-specific chronic inflammation. The International Journal of Biochemistry and Cell Biology (2009), doi:10.1016/j.biocel.2008.08.023
7. Bak, A., Brown, R., Minian, G., Porter, T.: Global actions, groupoid atlases and related topics. Journal of Homotopy and Related Structures 1, 1–54 (2006)
8. Bebbington, P.: Transcultural aspects of affective disorders. International Review of Psychiatry 5, 145–156 (1993)
9. Ben-Jonathan, N., Hugo, E., Brandenbourg, T.: Effects of bisphenol A on adipokine release from human adipose tissue: implications for the metabolic syndrome. Molecular Cell Endocrinology 304, 49–54 (2009)
10. Bennett, C.: Logical depth and physical complexity. In: Herkin, R. (ed.) The Universal Turing Machine: A Half-Century Survey, pp. 227–257. Oxford University Press, Oxford (1988)
11. Beran, J.: Statistics for Long-Memory Processes. Chapman and Hall, New York (1994)
12. Bos, R.: Continuous representations of groupoids. arXiv:math/0612639 (2007)
13. Bossdorf, O., Richards, C., Pigliucci, M.: Epigenetics for ecologists. Ecology Letters 11, 106–115 (2008)
14. Brown, G., Harris, T., Peto, J.: Life events and psychiatric disorders, II: nature of causal link. Psychological Medicine 3, 159–176 (1973)
15. Brown, R.: From groups to groupoids: a brief survey. Bulletin of the London Mathematical Society 19, 113–134 (1987)
16. Buneci, M.: Representare de Groupoizi. Editura Mirton, Timisoara (2003)

17. Burago, D., Burago, Y., Ivanov, S.: A Course in Metric Geometry. Graduate Studies in Mathematics, vol. 33. AMS, Providence (2001)
18. Cannas Da Silva, A., Weinstein, A.: Geometric Models for Noncommutative Algebras. AMS, RI (1999)
19. Chung, K., Williams, R.: Introduction to Stochastic Integration, 2nd edn. Birkhauser, Boston (1990)
20. Ciliberti, S., Martin, O., Wagner, A.: Robustness can evolve gradually in complex regulatory networks with varying topology. PLoS Computational Biology 3(2), e15 (2007)
21. Ciliberti, S., Martin, O., Wagner, A.: Innovation and robustness in complex regulatory gene networks. Proceeding of the National Academy of Sciences 104, 13591–13596 (2007)
22. Clougherty, J., Levy, J., Kubzansky, L., Ryan, P., Suglia, S., Canner, M., Wright, R.: Synergistic effects of traffic-related air pollution and exposure to violence on urban asthma etiology. Environmental Health Perspectives 115, 1140–1146 (2007)
23. Cohen, I.: Immune system computation and the immunological homunculus. In: Wang, J., Whittle, J., Harel, D., Reggio, G. (eds.) MoDELS 2006. LNCS, vol. 4199, pp. 499–512. Springer, Heidelberg (2006)
24. Cohen, I., Harel, D.: Explaining a complex living system: dynamics, multi-scaling, and emergence. Journal of the Royal Society: Interface 4, 175–182 (2007)
25. Cover, T., Thomas, J.: Elements of Information Theory. John Wiley and Sons, New York (1991)
26. de Groot, S., Mazur, R.: Non-Equilibrium Thermodynamics. Dover Publications, New York (1984)
27. Dehaene, S., Naccache, L.: Towards a cognitive neuroscience of consciousness: basic evidence and a workspace framework. Cognition 79, 1–37 (2001)
28. del Hoyo, M., Minian, E.: Classical invariants for global actions and groupoid atlases. Applied Categorical Structures 18, 689–721 (2008)
29. Dembo, A., Zeitouni, O.: Large Deviations: Techniques and Applications, 2nd edn. Springer, New York (1998)
30. Dietrich, K., Douglas, R., Succop, P., Berger, O., Bornschein, R.: Early exposure to lead and juvenile delinquency. Neurotoxicology and Teratology 23, 511–518 (2001)
31. Dohrenwend, B.P., Dohrenwend, B.S.: Social and cultural influences on psychopathology. Annual Review of Psychology 25, 417–452 (1974)
32. Dretske, F.: The explanatory role of information. Philosophical Transactions of the Royal Society A 349, 59–70 (1994)
33. Durham, W.: Coevolution: Genes, Culture and Human Diversity. Stanford University Press, Palo Alto (1991)
34. DSMIV: Diagnostic and Statistical Manual, 4th edn. American Psychiatric Association, Washington (1994)
35. Eaton, W.: Life events, social supports, and psychiatric symptoms: a re-analysis of the New Haven data. Journal of Health and Social Behavior 19, 230–234 (1978)
36. Ellis, R.: Entropy, Large Deviations, and Statistical Mechanics. Springer, New York (1985)
37. Emery, M.: Stochastic Calculus on Manifolds. Springer, New York (1989)
38. Feynman, R.: Lectures on Computation. Westview Press, New York (2000)
39. Foley, D., Craid, J., Morley, R., Olsson, C., Dwyer, T., Smith, K., Saffery, R.: Prospects for epigenetic epidemiology. American Journal of Epidemiology 169, 389–400 (2009)
40. Frankel, T.: The Geometry of Physics: An Introduction, 2nd edn. Cambridge University Press, Cambridge (2006)

41. Fullilove, M.: Root Shock: How Tearing Up City Neighborhoods Hurts America and What we can do about it. Balantine Books, New York (2004)
42. Gilbert, P.: Evolutionary approaches to psychopathology: the role of natural defenses. Austrailian and New Zealand Journal of Psychiatry 35, 17–27 (2001)
43. Gilbert, S.: Mechanisms for the environmental regulation of gene expression: ecological aspects of animal development. Journal of Bioscience 30, 65–74 (2001)
44. Glass, T., Bandeen-Roche, K., McAtee, M., Bolla, K., Todd, A., Schwartz, B.: Neighborhood psychosocial hazards and the association of cumulative lead dose with cognitive function in older adults. American Journal of Epidemiology 169, 683–692 (2009)
45. Glazebrook, J.F., Wallace, R.: Small worlds and red queens in the global workspace: an information-theoretic approach. Cognitive Systems Reserch 10, 333–365 (2009)
46. Glazebrook, J.F., Wallace, R.: Rate distortion manifolds as model spaces for cognitive information. Informatica Overview Article 33, 309–345 (2009)
47. Golubitsky, M., Stewart, I.: Nonlinear dynamics and networks: the groupoid formalism. Bulletin of the American Mathematical Society 43, 305–364 (2006)
48. Gunderson, L.: Ecological resilience – in theory and application. Annual Reviews of Ecological Systematics 31, 425–439 (2000)
49. Heine, S.: Self as cultural product: an examination of East Asian and North American selves. Journal of Personality 69, 881–906 (2001)
50. Helgason, S.: Differential Geometry and Symmetric Spaces. Academic Press, New York (1962)
51. Henrich J., Heine, S., Norenzayan, A.: The Weirdest people in the world? Behavioral and Brain Sciences (2010) (in press)
52. Holling, C.: Resilience and stability of ecological systems. Annual Reviews of Ecological Systematics 4, 1–23 (1973)
53. Hoppensteadt, F., Izhikevich, E.: Weakly Connected Neural Networks. Springer, New York (1997)
54. Jablonka, E., Lamb, M.: Epigenetic Inheritance and Evolution: The Lamarckian Dimension. Oxford University Press, Oxford (1995)
55. Jablonka, E., Lamb, M.: Epigenetic inheritance in evolution. Journal of Evolutionary Biology 11, 159–183 (1998)
56. Jablonka, E.: Epigenetic epidemiology. International Journal of Epidemiology 33, 929–935 (2004)
57. Jaenisch, R., Bird, A.: Epigenetic regulation of gene expression: how the genome integrates intrinsic and environmental signals. Nature Genetics Supplement 33, 245–254 (2003)
58. Jacobson, J., Jacobson, S.: Breast-feeding and gender as moderators of teratogenic effects on cognitive development. Neurotoxicological Teratology 24, 349–358 (2002)
59. Jenkins, J., Kleinman, A., Good, B.: Cross-cultural studies of depression. In: Becker, J., Kleinman, A. (eds.) Advances in mood disorders: Theory and Research, pp. 67–99. L. Erlbaum, Los Angeles (1990)
60. Johnson-Laird, P., Mancini, F., Gangemi, A.: A hyper-emotion theory of psychological illnesses. Psychological Reviews 113, 822–841 (2006)
61. Karp, R., Chen, C., Meyers, A.: The appearance of discretionary income: influence on the prevalence of under- and over-nutrition. International Journal of Equity in Health 4, 10 (2005)
62. Khinchin, A.: Mathematical Foundations of Information Theory. Dover, New York (1957)

63. Kleinman, A., Good, B.: Culture and Depression: Studies in the Anthropology of Cross-Cultural Psychiatry of Affect and Depression. University of California Press, Berkeley (1985)
64. Kleinman, A., Cohen, A.: Psychiatry's global challenge. Scientific American 276(3), 86–89 (1997)
65. Landau, L., Lifshitz, E.: Statistical Physics, Part I, 3rd edn. Elsevier, New York (2007)
66. Manolio, T., Collins, F., Cox, N., et al.: Finding the missing heritability of complex diseases. Nature 461, 747–753 (2009)
67. Manson, S.: Culture and major depression: Current challenges in the diagnosis of mood disorders. Psychiatric Clinics of North America 18, 487–501 (1995)
68. Markus, H., Kitayama, S.: Culture and the self- implications for cognition, emotion, and motivation. Psychological Review 98, 224–253 (1991)
69. Matsuda, T., Nisbett, R.: Culture and change blindness. Cognitive Science: A Multidisciplinary Journal 30, 381–399 (2006)
70. Maturana, H., Varela, F.: Autopoiesis and Cognition. Reidel Publishing Company, Dordrecht (1980)
71. Maturana, H., Varela, F.: The Tree of Knowledge. Shambhala Publications, Boston (1992)
72. Marsella, A.: Cultural aspects of depressive experience and disorders. In: Lonner, W., Dinnel, D., Hays, S., Sattler, D. (eds.) Online Readings in Psychology and Culture (Unit 9, ch. 4), Center for Cross-Cultural Research, Western Washington University, Bellingham, WA (2003), http://www.wwu.edu/~culture
73. Miranda, M., Kim, D., Overstreet Galeano, M., Paul, C., Hull, A., Morgan, S.: The relationship between early childhood blood lead levels and performance on end-of-grade tests. Environmental Health Perspectives 115, 1242–1247 (2007)
74. Needleman, H., Riess, J., Tobin, M., Biesecker, G., Greenhouse, J.: Bone lead levels and delinquent behavior. Journal of the American Medical Association 275, 363–369 (1996)
75. Nesse, R.: Is depression an adaptation? Archives of General Psychiatry 57, 14–20 (2000)
76. Nisbett, R., Peng, K., Incheol, C., Norenzayan, A.: Culture and systems of thought: holistic vs. analytic cognition. Psychological Review 108, 291–310 (2001)
77. Norenzayan, A., Heine, S.: Psychological universals: what are they and how can we know? Psychological Bulletin 131, 763–784 (2005)
78. O'Nuallain, S.: Code and context in gene expression, cognition, and consciousness. In: Barbiere, M. (ed.) The Codes of Life: The Rules of Macroevolution, ch. 15, pp. 347–356. Springer, New York (2008)
79. Pettini, M.: Geometry and Topology in Hamiltonian Dynamics and Statistical Mechanics. Springer, New York (2007)
80. Prandolini, R., Moody, M.: Brownian nature of the Time-Base Error in tape recordings. Journal of the Audio Engineering Society 43, 241–247 (1995)
81. Protter, P.: Stochastic Integration and Differential Equations: A New Approach. Springer, New York (1995)
82. Richerson, P., Boyd, R.: Not by Genes Alone: How Culture Transformed Human Evolution. Chicago University Press, Chicago (2004)
83. Rockafellar, R.: Complex Analysis. Princeton University Press, Princeton (1970)
84. Sarshar, N., Wu, X.: On Rate-Distortion models for natural images and wavelet coding performance. IEEE Transactions on Image Processing 16, 1383–1394 (2007)

85. Sarlio-Lahteenkorva, S., Lahelma, E.: Food insecurity is associated with past and present economic disadvantage and body mass index. Journal of Nutrition 131, 2880–2884 (2001)
86. Scherrer, K., Jost, J.: The gene and the genon concept: a functional and information-theoretic analysis. Molecular Systems Biology 3, 87–93 (2007)
87. Scherrer, K., Jost, J.: Gene and genon concept: coding versus regulation. Theory in Bioscience 126, 65–113 (2007)
88. Shankardass, K., McConnell, R., Jerrett, M., Milam, J., Richardson, J., Berhane, K.: Parental stress increases the effect of traffic-related air pollution on childhood asthma incidence. Proceedings of the National Academy of Sciences 106, 12406–12411 (2009)
89. Skierski, M., Grundland, A., Tuszynski, J.: Analysis of the three-dimensional time-dependent Landau-Ginzburg equation and its solutions. Journal of Physics A (Math. Gen.) 22, 3789–3808 (1989)
90. Toulouse, G., Dehaene, S., Changeux, J.: Spin glass model of learning by selection. Proceedings of the National Academy of Sciences 83, 1695–1698 (1986)
91. Turner, B.: Histone acetylation and an epigeneticv code. Bioessays 22, 836–845 (2000)
92. Wallace, D., Wallace, R.: Scales of geography, time, and population: the study of violence as a public health problem. American Journal of Public Health 88, 1853–1858 (1998)
93. Wallace, D., Wallace, R.: Life and death in Upper Manhattan and the Bronx: Toward evolutionary perspectives on catastrophic social change. Environment and Planning A 32, 1245–1266 (2000)
94. Wallace, D., Wallace, R., Rauh, V.: Community stress, demoralization, and body mass index: evidence for social signal transduction. Social Science and Medicine 56, 2467–2478 (2003)
95. Wallace, R.: Consciousness: A Mathematical Treatment of the Global Neuronal Workspace Model. Springer, New York (2005)
96. Wallace, R.: Culture and inattentional blindness. Journal of Theoretical Biology 245, 378–390 (2007)
97. Wallace, R.: Developmental disorders as pathological resilience domains. Ecology and Society 13, 29 (2008) (online)
98. Wallace, R., Fullilove, M.: Collective Consciousness and its Discontents: Institutional Distributed Cognition, Racial Policy, and Public Health in the United States. Springer, New York (2008)
99. Wallace, R., Wallace, D.: Structured psychosocial stress and the US obesity epidemic. Journal of Biological Systems 13(2005), 363–384 (2005)
100. Wallace, R., Wallace, D.: Punctuated equilibrium in statistical models of generalized coevolutionary resilience: How sudden ecosystem transitions can entrain both phenotype expression and darwinian selection. In: Priami, C. (ed.) Transactions on Computational Systems Biology IX. LNCS (LNBI), vol. 5121, pp. 23–85. Springer, Heidelberg (2008)
101. Wallace, R., Wallace, D.: Code, context, and epigenetic catalysis in gene expression. In: Priami, C., Back, R.-J., Petre, I. (eds.) Transactions on Computational Systems Biology XI. LNCS, vol. 5750, pp. 283–334. Springer, Heidelberg (2009)
102. Wallace, R.G., Wallace, R.: Evolutionary radiation and the spectrum of consciousness. Consciousness and Cognition 18, 160–167 (2009)
103. Wang, M., Uhlenbeck, G.: On the theory of the Brownian Motion II. Reviews of Modern Physics 17, 323–342 (1945)

104. Weinstein, A.: Groupoids: unifying internal and external symmetry. Notices of the American Mathematical Association 43, 744–752 (1996)
105. West-Eberhard, M.: Developmental Placisticity and Evolution. Oxford University Press, New York (2003)
106. West-Eberhard, M.: Developmental plasticity and the origin of species differences. Proceedings of the National Academy of Sciences 102, 6543–6549 (2005)
107. Zhu, R., Rebirio, A., Salahub, D., Kaufmann, S.: Studying genetic regulatory networks at the molecular level: delayed reaction stochastic models. Journal of Theoretical Biology 246, 725–745 (2007)

A Mathematical Appendix: Groupoids

A.1 Basic Ideas

Following [104] closely, a groupoid, G, is defined by a base set A upon which some mapping – a morphism – can be defined. Note that not all possible pairs of states (a_j, a_k) in the base set A can be connected by such a morphism. Those that can define the groupoid element, a morphism $g = (a_j, a_k)$ having the natural inverse $g^{-1} = (a_k, a_j)$. Given such a pairing, it is possible to define 'natural' end-point maps $\alpha(g) = a_j, \beta(g) = a_k$ from the set of morphisms G into A, and a formally associative product in the groupoid $g_1 g_2$ provided $\alpha(g_1 g_2) = \alpha(g_1), \beta(g_1 g_2) = \beta(g_2)$, and $\beta(g_1) = \alpha(g_2)$. Then the product is defined, and associative, $(g_1 g_2)g_3 = g_1(g_2 g_3)$.

In addition, there are natural left and right identity elements λ_g, ρ_g such that $\lambda_g g = g = g \rho_g$ [104].

An orbit of the groupoid G over A is an equivalence class for the relation $a_j \sim G a_k$ if and only if there is a groupoid element g with $\alpha(g) = a_j$ and $\beta(g) = a_k$. Following [18], we note that a groupoid is called transitive if it has just one orbit. The transitive groupoids are the building blocks of groupoids in that there is a natural decomposition of the base space of a general groupoid into orbits. Over each orbit there is a transitive groupoid, and the disjoint union of these transitive groupoids is the original groupoid. Conversely, the disjoint union of groupoids is itself a groupoid.

The isotropy group of $a \in X$ consists of those g in G with $\alpha(g) = a = \beta(g)$. These groups prove fundamental to classifying groupoids.

If G is any groupoid over A, the map $(\alpha, \beta) : G \to A \times A$ is a morphism from G to the pair groupoid of A. The image of (α, β) is the orbit equivalence relation $\sim G$, and the functional kernel is the union of the isotropy groups. If $f : X \to Y$ is a function, then the kernel of f, $ker(f) = [(x_1, x_2) \in X \times X : f(x_1) = f(x_2)]$ defines an equivalence relation.

Groupoids may have additional structure. As Weinstein [104] explains, a groupoid G is a topological groupoid over a base space X if G and X are topological spaces and α, β and multiplication are continuous maps. A criticism sometimes applied to groupoid theory is that their classification up to isomorphism is nothing other than the classification of equivalence relations via the orbit equivalence relation and groups via the isotropy groups. The imposition of a compatible topological structure produces a nontrivial interaction between the two structures.

In essence, a groupoid is a category in which all morphisms have an inverse, here defined in terms of connection to a base point by a meaningful path of an information source dual to a cognitive process.

As [104] points out, the morphism (α, β) suggests another way of looking at groupoids. A groupoid over A identifies not only which elements of A are equivalent to one another (isomorphic), but *it also parametizes the different ways (isomorphisms) in which two elements can be equivalent*, i.e., all possible information sources dual to some cognitive process. Given the information theoretic characterization of cognition presented above, this produces a full modular cognitive network in a highly natural manner.

Brown [15] describes the fundamental structure as follows:

> A groupoid should be thought of as a group with many objects, or with many identities... A groupoid with one object is essentially just a group. So the notion of groupoid is an extension of that of groups. It gives an additional convenience, flexibility and range of applications...

Example 1. A disjoint union [of groups] $G = \cup_\lambda G_\lambda, \lambda \in \Lambda$, is a groupoid: the product ab is defined if and only if a, b belong to the same G_λ, and ab is then just the product in the group G_λ. There is an identity 1_λ for each $\lambda \in \Lambda$. The maps α, β coincide and map G_λ to λ, $\lambda \in \Lambda$.

Example 2. An equivalence relation R on [a set] X becomes a groupoid with $\alpha, \beta : R \to X$ the two projections, and product $(x, y)(y, z) = (x, z)$ whenever $(x, y), (y, z) \in R$. There is an identity, namely (x, x), for each $x \in X$...

Weinstein [104] makes the following fundamental point:

> Almost every interesting equivalence relation on a space B arises in a natural way as the orbit equivalence relation of some groupoid G over B. Instead of dealing directly with the orbit space B/G as an object in the category S_{map} of sets and mappings, one should consider instead the groupoid G itself as an object in the category G_{htp} of groupoids and homotopy classes of morphisms.

The groupoid approach has become quite popular in the study of networks of coupled dynamical systems which can be defined by differential equation models, (e.g., [47]).

A.2 Global and Local Symmetry Groupoids

Here we again follow [104] fairly closely. Consider a tiling of the euclidean plane R^2 by identical 2 by 1 rectangles, specified by the set X (one dimensional) where the grout between tiles is $X = H \cup V$, having $H = R \times Z$ and $V = 2Z \times R$, where R is the set of real numbers and Z the integers. Call each connected component of $R^2 \backslash X$, that is, the complement of the two dimensional real plane intersecting X, a tile.

Let Γ be the group of those rigid motions of R^2 which leave X invariant, i.e., the normal subgroup of translations by elements of the lattice $\Lambda = H \cap V = 2Z \times Z$ (corresponding to corner points of the tiles), together with reflections through each of the points $1/2\Lambda = Z \times 1/2Z$, and across the horizontal and vertical lines through those points. As noted in [104], much is lost in this coarse-graining, in particular the same symmetry group would arise if we replaced X entirely by the lattice Λ of corner points. Γ retains no information about the local structure of the tiled plane. In the case of a real tiling, restricted to the finite set $B = [0, 2m] \times [0, n]$ the symmetry group shrinks drastically: The subgroup leaving $X \cap B$ invariant contains just four elements even though a repetitive pattern is clearly visible. A two-stage groupoid approach recovers the lost structure.

We define the transformation groupoid of the action of Γ on R^2 to be the set

$$G(\Gamma, R^2) = \{(x, \gamma, y | x \in R^2, y \in R^2, \gamma \in \Gamma, x = \gamma y\},$$

with the partially defined binary operation

$$(x, \gamma, y)(y, \nu, z) = (x, \gamma\nu, z).$$

Here $\alpha(x, \gamma, y) = x$, and $\beta(x, \gamma, y) = y$, and the inverses are natural.

We can form the restriction of G to B (or any other subset of R^2) by defining

$$G(\Gamma, R^2)|_B = \{g \in G(\Gamma, R^2) | \alpha(g), \beta(g) \in B\}$$

(1). An orbit of the groupoid G over B is an equivalence class for the relation $x \sim_G y$ if and only if there is a groupoid element g with $\alpha(g) = x$ and $\beta(g) = y$.

Two points are in the same orbit if they are similarly placed within their tiles or within the grout pattern.

(2). The isotropy group of $x \in B$ consists of those g in G with $\alpha(g) = x = \beta(g)$. It is trivial for every point except those in $1/2\Lambda \cap B$, for which it is $Z_2 \times Z_2$, the direct product of integers modulo two with itself.

By contrast, embedding the tiled structure within a larger context permits definition of a much richer structure, i.e., the identification of local symmetries.

We construct a second groupoid as follows. Consider the plane R^2 as being decomposed as the disjoint union of $P_1 = B \cap X$ (the grout), $P_2 = B \backslash P_1$ (the complement of P_1 in B, which is the tiles), and $P_3 = R^2 \backslash B$ (the exterior of the tiled room). Let E be the group of all euclidean motions of the plane, and define the local symmetry groupoid G_{loc} as the set of triples (x, γ, y) in $B \times E \times B$ for which $x = \gamma y$, and for which y has a neighborhood $\mathcal{U}$ in R^2 such that $\gamma(\mathcal{U} \cap P_i) \subseteq P_i$ for $i = 1, 2, 3$. The composition is given by the same formula as for $G(\Gamma, R^2)$.

For this groupoid-in-context there are only a finite number of orbits:

$\mathcal{O}_1$ = interior points of the tiles.

$\mathcal{O}_2$ = interior edges of the tiles.

$\mathcal{O}_3$ = interior crossing points of the grout.

$\mathcal{O}_4$ = exterior boundary edge points of the tile grout.

$\mathcal{O}_5$ = boundary 'T' points.

$\mathcal{O}_6$ = boundary corner points.

The isotropy group structure is, however, now very rich indeed:

The isotropy group of a point in $\mathcal{O}_1$ is now isomorphic to the entire rotation group O_2.

It is $Z_2 \times Z_2$ for $\mathcal{O}_2$.

For $\mathcal{O}_3$ it is the eight-element dihedral group D_4.

For $\mathcal{O}_4, \mathcal{O}_5$ and $\mathcal{O}_6$ it is simply Z_2.

These are the 'local symmetries' of the tile-in-context.

Refining Dynamics of Gene Regulatory Networks in a Stochastic π-Calculus Framework

Loïc Paulevé, Morgan Magnin, and Olivier Roux

IRCCyN, UMR CNRS 6597,
École Centrale de Nantes, France
{loic.pauleve,morgan.magnin,olivier.roux}@irccyn.ec-nantes.fr

Abstract. In this paper, we introduce a framework allowing to model and analyse efficiently Gene Regulatory Networks (GRNs) in their temporal and stochastic aspects. The analysis of stable states and inference of René Thomas' discrete parameters derives from this logical formalism. We offer a compositional approach which comes with a natural translation to the Stochastic π-Calculus. The method we propose consists in successive refinements of generalised dynamics of GRNs. We illustrate the merits and scalability of our framework on the control of the differentiation in a GRN generalising metazoan segmentation processes, and on the analysis of stable states within a large GRN studied in the scope of breast cancer researches.

1 Introduction

Modelling, analysis and numerical or stochastic simulations are a usual means to predict the behaviour of complex living systems such as interacting genes.

Regulations between genes (activation or inhibition) are generally represented by Gene Regulatory Network (GRN) graphs. However, a GRN graph is not enough to describe dynamics. In continuous frameworks such as ordinary differential equations, parameters for differential equations are needed. In logical (or qualitative) frameworks such as boolean or discrete networks, dynamics are driven by René Thomas' parameters or equivalent [1].

Hybrid modelling brings quantitative aspects — such as temporal or stochastic parameters — to logical modelling. In the field of formal languages, κ language [2] or Stochastic π-Calculus [3,4,5] bring theoretical Computer Science frameworks for biological modelling. In the field of formal verifications of biological systems, frameworks like Time or Stochastic Petri Nets [6,7], Biocham [8], Timed Automata [9], Charon (Hybrid Modelling) [10] and Linear Hybrid Modelling [11] bring the first bricks for verifying and controlling dynamics of such systems.

Inference of temporal and stochastic parameters is still challenging as the domain of parameters is continuous and its volume generally grows exponentially with the number of genes. Compositional approaches, inherent to process algebras, aspire at reducing this complexity by allowing a local reasoning.

Our aim is temporal parameters synthesis for verifying formal properties on hybrid models.

C. Priami et al. (Eds.): Trans. on Comput. Syst. Biol. XIII, LNBI 6575, pp. 171–191, 2011.
© Springer-Verlag Berlin Heidelberg 2011

Our contribution consists in the introduction of both temporal and stochastic parameters into process algebra models of GRNs through a new Stochastic π-Calculus framework: the Process Hitting framework.

Starting from a GRN without any other parameters, its largest dynamics are expressed in Process Hitting and then are refined to match the expected behaviour. Such a refinement is achieved by constructing cooperativity between genes and by creating stable states. As we will show, detecting stable states of a Process Hitting is straightforward, as well as infering the René Thomas' discrete parameters K.

Moreover, the Stochastic π-Calculus naturally brings time and stochasticity into our Process Hitting framework. We introduce a *stochasticity absorption factor* to highlight either the temporal or stochastic aspect of reactions. The direct translation of Process Hitting to the Stochastic π-Calculus allows simulations of such models by softwares like BioSpi [3] or SPiM [12].

Several points motivate the choice of the Stochastic π-Calculus framework for introducing the Process Hitting. The Stochastic π-Calculus can be considered as a "low-level" process algebra: there are very few operators compared, for instance, to the Beta Workbench [13] or Bio-PEPA [14]. This makes the presentation of the Process Hitting as a Stochastic π-Calculus framework more generic. Moreover, the Stochastic π-Calculus comes with a bunch of established tools as previously cited and translations into various framework have already been formalized, such as to PRISM, a probabilistic model checking tool [15,16].

This paper is structured as follows. Section 2 introduces our framework and how it is used to build the generalised dynamics of a GRN. Section 3 presents dynamics refinement techniques and Section 4 shows how infering the René Thomas' parameters leading to such dynamics. Introduction of both temporal and stochastic parameters within Process Hitting models is addressed in Section 5. The overall approach is illustrated by two applications in Section 6. The first applies the refinement method to a toy GRN involved in biological segmentations phenomena. The temporal and stochastic parameters are then infered to bring a particular behaviour to the system. The second application shows the scalability of the Process Hitting framework by modelling a large GRN composed of 20 genes.

Notations. Given a set S, $\overbrace{S \times \cdots \times S}^{n}$, will be abbreviated as S^n. If S is finite and countable, we note $|S|$ its cardinality. Given a n-tuple C, $C[x/y]$ refers to the n-tuple C within the element y has been substituted by x. Belonging and cartesian product for n-tuples are defined similarly to sets. $[x_i; x_{i+n}]$ refers to the interval $\{x_i, x_{i+1}, \ldots, x_{i+n}\}$. '$\wedge$' stands for the logical *and* connector.

2   Generalised Dynamics for Gene Regulatory Networks

First, we recall the basis of the René Thomas' discrete modelling framework from which we designed our refinement approach. This method is described in subsections 2.2 and 2.3. This leads to a straightforward translation into the π-Calculus which makes it possible to express generalised dynamics of GRNs.

2.1   Gene Regulatory Networks

GRNs are often described by interaction graphs where nodes are genes with activation and inhibition relations respectively represented by positive and negative edges [1,17].

In the discrete framework of René Thomas, each gene has at least two qualitative levels. The influence of an activating (resp. inhibiting) gene on its target depends on a threshold value: when the level of the gene is greater or equal than the threshold, the gene holds a positive (resp. negative) effect; when the level of the gene is lower than the threshold, the gene holds a negative (resp. positive) effect [1].

Definition 1 (Gene Regulatory Network Graph). *A* Gene Regulatory Network graph *is a triple* (Γ, E_+, E_-) *where* Γ *is the finite set of genes and* $a \xrightarrow{t} b \in E_+$ *(resp. E_-), t positive integer, if and only if the gene a above level t is an activator (resp. inhibitor) for b. We note a_i the level i of the gene a.*

Given a GRN graph (Γ, E_+, E_-), the maximum qualitative level for gene $a \in \Gamma$ is noted a_{l_a} where l_a is the highest threshold involved in its regulations:

$$l_a = max(\{t \mid \exists b \in \Gamma, a \xrightarrow{t} b \in E_+ \cup E_-\}) \ . \tag{1}$$

We denote $levels_+(a, b)$ (resp. $levels_-(a, b)$) the set of levels of a where a effectively activates (resp. inhibits) b.

Definition 2 (Effective Levels). *If $a \xrightarrow{t} b \in E_+$, $levels_+(a, b) = [a_t; a_{l_a}]$ and $levels_-(a, b) = [a_0; a_{t-1}]$. If $a \xrightarrow{t} b \in E_-$, $levels_+(a, b) = [a_0; a_{t-1}]$ and $levels_-(a, b) = [a_t; a_{l_a}]$. Else $levels_+(a, b) = levels_-(a, b) = \emptyset$.*

2.2   The Process Hitting Framework

We want to describe the action of a gene at a given level on another one. If the gene a at a given level i is an activator for b, it has a positive action on b, meaning the level of b will tend to increase. Conversely if a at a level i' is an inhibitor for b, it has a negative action on b and then the level of b will tend to decrease.

The action is "a at level i making b at level j increase (or decrease) to level k". We say a_i hits b_j to make it *bounce* to b_k and note this action $a_i \rightarrow b_j \,\Gamma\, b_k$. In the process hitting framework, a_i, b_j, b_k are refered as *processes* and a, b as *sorts*. Sorts can represent genes, but also logical entities, as described in further sections.

Definition 3 (Action). *An action is noted $a_i \rightarrow b_j \,\Gamma\, b_k$ where a_i is a process of sort a and $b_j \neq b_k$ two processes of sort b. $a_i \rightarrow b_j$ is the* hit *part, and $b_j \,\Gamma\, b_k$ the* bounce *part. When $a_i = b_j$, such an action is refered as a self-action and a_i is called a* self-hitting process.

In this paper, one and only one process of each sort is present at any instant. Hence, hits between different processes of a same sort are prohibited. The set of these living processes gives the state of the Process Hitting.

Definition 4 (Process Hitting). *A Process Hitting* $\mathcal{PH}$ *is a triple* $(\Sigma, L, \mathcal{H})$:

- $\Sigma = \{a, b, \dots\}$ *is the finite countable set of sorts,*
- $L = \prod_{a \in \Sigma} L_a$ *is the set of states for* $\mathcal{PH}$, *with* $L_a = \{a_0 \dots a_{l_a}\}$ *the finite and countable set of processes of sort* $a \in \Sigma$ *and* l_a *a positive integer,* $a \neq b \Rightarrow a_i \neq b_j \ \forall (a_i, b_j) \in L_a \times L_b$,
- $\mathcal{H} = \{a_i \rightarrow b_j \ \text{⌐} \ b_k, \cdots \mid (a, b) \in \Sigma^2, \ (a_i, b_j, b_k) \in L_a \times L_b \times L_b, b_j \neq b_k, a = b \Rightarrow a_i = b_j\}$, *is the finite set of actions.*

At a given state $s \in L$, an action $a_i \rightarrow b_j \ \text{⌐} \ b_k$ is playable if both processes a_i and b_j are present in s. When this action is played, the process b_k replaces b_j.

Definition 5 (Next States). *Let* $(\Sigma, L, \mathcal{H})$ *be a Process Hitting and* $s \in L$ *be one of its states. The set of the next possible states for* s *are computed as follows:*

$$next(s) = \{s[b_k/b_j] \mid \exists(a_i, b_j) \in s^2, \exists b_k \in L_b, a_i \rightarrow b_j \ \text{⌐} \ b_k \in \mathcal{H}\} \ .$$

Definition 6 (Stable state). *Let* $\mathcal{PH} = (\Sigma, L, \mathcal{H})$ *be a Process Hitting and* $s \in L$ *be a state,* s *is a* stable state *for* $\mathcal{PH}$ *if and only if* $next(s) = \emptyset$.

2.3 Graphical Representations of a Process Hitting

We set up two complementary graphical representations of a Process Hitting. The first one exhibits the actions between process levels, the second one points out the absence of hits between them. We finally define the State Graph of a Process Hitting. Figure 1 shows an instance for each of these three representations.

Given a Process Hitting $(\Sigma, L, \mathcal{H})$, its *Hypergraph* represents each action $a_i \rightarrow b_j \ \text{⌐} \ b_k \in \mathcal{H}$ by a directed hyperedge from a_i to b_k passing by b_j. The hit part (a_i to b_j) is drawn as a plain edge and the bounce part (b_j to b_k) as a dotted edge.

Definition 7 (Process Hitting Hypergraph). *The* Hypergraph *of a Process Hitting* $(\Sigma, L, \mathcal{H})$ *is a couple* (P, A) *where* $P = \bigcup_{a \in \Sigma} L_a$ *are the vertices and* $A \subseteq P^3$ *the directed hyperedges given by* $A = \{(a_i, b_j, b_k) \mid a_i \rightarrow b_j \ \text{⌐} \ b_k \in \mathcal{H}\}$.

In the following we introduce a complementary representation we call *Hitless Graph*. It will allow us to obtain extra results such as the stable states of a Process Hitting (Section 3.2). The Hitless Graph of a Process Hitting $(\Sigma, L, \mathcal{H})$ relates two processes of different sorts if and only if they hit neither each other nor themselves. Vertices of a Hitless Graph may be split into $n \leq |\Sigma|$ partitions having no element inside related to each other: a partition is, for any sort, a subset of its processes without self-actions. Such a graph is called *n-partite*.

Definition 8 (n-Partite Graph). *A graph* $G = (V, E)$ *is* n-partite *if and only if* $V = \bigcup_{k=1}^{n} V_k$, $V_k \neq \emptyset$, $\forall 1 \leq k, k' \leq n, V_k \cap V_{k'} = \emptyset$ *and* $(a_i, b_j) \in E \Rightarrow \exists 1 \leq k \neq k' \leq n$, $a_i \in V_k \wedge b_j \in V_{k'}$.

Definition 9 (Hitless Graph). *Given a Process Hitting* $\mathcal{PH} = (\Sigma, L, \mathcal{H})$, *its Hitless Graph* $\overline{\mathcal{PH}} = (V, E)$ *is defined as a non-directed graph where the vertices* V *and edges* E *are computed as follows:*

$$V = \bigcup_{a \in \Sigma} \{a_i \in L_a \mid \forall a_{i'} \in L_a \nexists\, a_i \to a_i \upharpoonright a_{i'} \in \mathcal{H}\}$$

$$E = \{(a_i, b_j) \in V^2 \mid \forall b_{j'} \in L_b \nexists\, a_i \to b_j \upharpoonright b_{j'} \in \mathcal{H}$$
$$\wedge\ \forall a_{i'} \in L_a \nexists\, b_j \to a_i \upharpoonright a_{i'} \in \mathcal{H}\} \ .$$

Property 1. By construction of V and E, $\overline{\mathcal{PH}}$ is a n-partite graph, $n \leq |\Sigma|$, where each partition is a subset of processes for one and only one sort and to each sort corresponds at most one partition.

We also define the n-cliques of a graph which are subsets of n vertices such that each element is related to each other.

Definition 10 (n-Clique). *Given a graph* $G = (V, E)$, $C \subseteq V$ *is a* $|C|$-clique *of* G *if and only if* $\forall (a_i, b_j) \in C^2, \{a_i, b_j\} \in E$.

Property 2. n-cliques of a n-partite graph have one and only one vertex in each partition.

Finally, the *State Graph* of a Process Hitting represents the possible transitions between each couple of its states.

Definition 11 (State Graph). *Given a Process Hitting* $(\Sigma, L, \mathcal{H})$, *its* State Graph *is a directed graph* $\mathcal{S} = (L, E \subseteq L^2)$ *with* $(s, s') \in E \Leftrightarrow s' \in next(s)$.

2.4 From Process Hitting to the π-Calculus

A main advantage of our approach is its natural translation to the π-Calculus. In this subsection we propose a method to translate any Process Hitting into a π-Calculus expression.

We briefly present the fragment of the π-Calculus which is sufficient for translating a Process Hitting. The full syntax for π-Calculus and examples can be found in [5,18]. π-Calculus expressions compose two kinds of objects: independently defined processes and channels shared by some processes. A process P has the capability to output (resp. input) on a channel γ and then become P', noted $!\gamma.P'$ (resp. $?\gamma.P'$). Output and input are synchronized operations, i.e. an outputting process is blocked until another process inputs on the same channel. A process can also execute an internal action (τ), nil operation $(\mathbf{0})$ or one amongst several $(P' + P'')$.

Let $\mathcal{PH} = (\Sigma, L, \mathcal{H})$ be a Process Hitting. For each process a_i of $\mathcal{PH}$, a π-Calculus process A_i is defined as follows. For each action $a_i \to b_j \upharpoonright b_k \in \mathcal{H}$ where

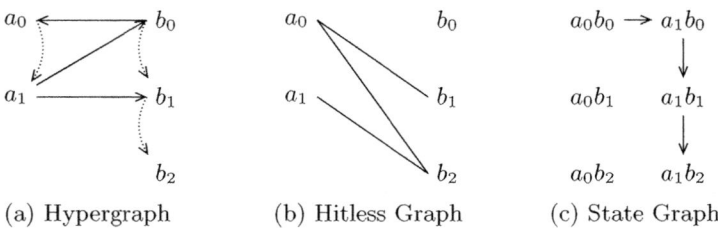

(a) Hypergraph   (b) Hitless Graph   (c) State Graph

Fig. 1. Graphical representations for the Process Hitting $\mathcal{PH} = (\{a, b\}, \{a_0, a_1\} \times \{b_0, b_1, b_2\}, \mathcal{H})$ with $\mathcal{H} = \{b_0 \rightarrow a_0 \restriction a_1, a_1 \rightarrow b_0 \restriction b_1, a_1 \rightarrow b_1 \restriction b_2\}$.

$a \neq b$, a new channel γ_α is created. The π-Calculus process A_i has the ability to output on this channel and the π-Calculus process B_j has the ability to input on this channel so as to become B_k (2). For each self-action $a_i \rightarrow a_i \restriction a_j \in \mathcal{H}$, A_i has the ability to become A_j after performing an internal action τ_α (3).

$$A_i ::= \sum_{\substack{\alpha = a_i \rightarrow b_j \restriction b_k \in \mathcal{H} \\ a \neq b}} !\gamma_\alpha.A_i + \sum_{\substack{\alpha = b_j \rightarrow a_i \restriction a_k \in \mathcal{H} \\ a \neq b}} ?\gamma_\alpha.A_k \qquad (2)$$

$$+ \sum_{\alpha = a_i \rightarrow a_i \restriction a_k \in \mathcal{H}} \tau_\alpha.A_k \qquad (3)$$

2.5 Generalised Dynamics for Gene Regulatory Networks

Our method to analyse GRNs takes benefit from the use of refinement techniques. Starting from the largest set of possible dynamics for the GRN, we gradually take into account only the specified behaviours and exclude the other ones, thus leading to a restrictive process.

We call this largest set of dynamics the *generalised dynamics* for the GRN graph. It is described by the following rules: the level of a gene increases (resp. decreases) if and only if at least one of its activators (resp. inhibitors) is present. The absence of activators is equivalent to the presence of one inhibitor.

Let $\mathcal{G} = (\Gamma, E_+, E_-)$ be a GRN graph. For all $(a, b) \in \Gamma^2$, we build the set of actions $\mathcal{H}_a^b$ from a to b reflecting the rules above:

– If $a \xrightarrow{t} b \in E_+$, all processes of sort a below the threshold t hit all processes of sort b but b_0 to make them decrease to the level below. Moreover, all processes of sort a above the threshold t hit all processes of sort b but b_{l_b} to make them increase to the level above:

$$\mathcal{H}_a^b = \{a_i \rightarrow b_j \restriction b_{j-1} \mid 0 \leq i < t, 1 \leq j \leq l_b\}$$
$$\cup \{a_{i'} \rightarrow b_{j'} \restriction b_{j'+1} \mid t \leq i' \leq l_a, 0 \leq j' < l_b\} .$$

– If $a \xrightarrow{t} b \in E_-$, the actions are defined similarly to the previous case except for the bounce directions which are reversed:

$$\mathcal{H}_a^b = \{a_i \rightarrow b_j \upharpoonright b_{j+1} \mid 0 \leq i < t, 0 \leq j < l_b\}$$
$$\cup \{a_{i'} \rightarrow b_{j'} \upharpoonright b_{j'-1} \mid t \leq i' \leq l_a, 1 \leq j' \leq l_b\} \ .$$

– If $b = a$ and $\nexists c \in \Gamma$, $c \xrightarrow{t} b \in E_- \cup E_+$, gene b lives in absence of activators: all processes of sort b but b_0 hit themselves to decrease to the level below.

$$\mathcal{H}_b^b = \{b_i \rightarrow b_i \upharpoonright b_{i-1} \mid 1 \leq i \leq l_b\} \ .$$

– Obviously, if $a \xrightarrow{t} b \notin E_- \cup E_+$ for any t and the previous case does not hold, we define $\mathcal{H}_a^b = \emptyset$.

The Process Hitting for the generalised dynamics of $\mathcal{G}$ is given by

$$\mathcal{PH} = \left(\Gamma, \ \prod_{a \in \Gamma}\{a_0, \ldots, a_{l_a}\}, \ \bigcup_{(a,b) \in \Gamma^2} \mathcal{H}_a^b\right) \ .$$

3   Refining Dynamics of Gene Regulatory Networks

We present two methods which aim at narrowing the set of dynamics of a Process Hitting for a GRN: the first one is based on cooperativity between genes, the other one deals with the knowledge of the stable states.

3.1   Cooperative Hits

Given two genes c and f regulating a gene a, the action of c on a may depend on the level of f: there exists a cooperativity between c and f on a. In discrete frameworks, the cooperativity is often described by a boolean function between genes levels [1,19]. We show how to build cooperativity within Process Hitting.

Let $(\Sigma, L, \mathcal{H})$ be a Process Hitting and $\sigma \subset \Sigma$ be a set of sorts cooperating on a given process a_k to make it bounce to $a_{k'}$. The set of all states of the cooperating sorts is denoted by $S = \prod_{z \in \sigma} L_z$. The subset of states where the cooperativity is effective is defined by $\top \subset S$.

For applying this cooperation, a new sort is added to the Process Hitting. This sort is called a *cooperative sort* and is refered as υ. The set of processes of sort υ is defined by $L_\upsilon = \{\upsilon_\varsigma, \forall \varsigma \in S\}$. Each process z_i of sort $z \in \sigma$ hits processes υ_ς of the cooperative sort υ where $z_i \notin \varsigma$ to make it bounce to $\upsilon_{\varsigma[z_i/z_j]}, z_j \in \varsigma$. We denote $\mathcal{H}_\sigma$ the set of such actions (4). In this way, the process of sort υ reflects the current state of its representatives.

The cooperativity between υ processes is added into the Process Hitting by replacing hits $\mathcal{H}_{coop}$ from processes of sorts present in σ to c_k (5) by hits $\mathcal{H}'_{coop}$ from processes of the cooperative sort υ selected in $\top$ (6).

$$\mathcal{H}_\sigma = \{z_i \rightarrow \upsilon_\varsigma \upharpoonright \upsilon_{\varsigma[z_i/z_j]} \mid \forall z \in \sigma, \forall (z_i, z_j) \in L_z^2, \forall \upsilon_\varsigma \in L_\upsilon, z_j \in \varsigma\} \quad (4)$$
$$\mathcal{H}_{coop} = \{z_i \rightarrow a_k \upharpoonright a_{k'} \in \mathcal{H} \mid \forall z \in \sigma\} \quad (5)$$
$$\mathcal{H}'_{coop} = \{\upsilon_\varsigma \rightarrow a_k \upharpoonright a_{k'} \mid \forall \varsigma \in \top\} \ . \quad (6)$$

The resulting Process Hitting is $(\Sigma \cup \{\upsilon\}, L \times L_\upsilon, (\mathcal{H} \setminus \mathcal{H}_{coop}) \cup \mathcal{H}_\sigma \cup \mathcal{H}'_{coop})$.

Example 1. Let $(\{f, c, a\}, \{f_0, f_1\} \times \{c_0, c_1\} \times \{a_0, a_1\}, \mathcal{H})$ be a Process Hitting where $\{f_1 \rightarrow a_0 \upharpoonright a_1, c_0 \rightarrow a_0 \upharpoonright a_1\} \subset \mathcal{H}$. The creation of a cooperativity between f_1 and c_0 on a_0 ($\sigma = \{f, c\}, \top = \{f_1 c_0\}$) is illustrated by Figure 2.

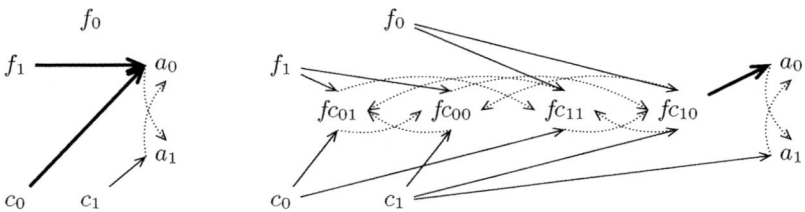

Fig. 2. Construction of a cooperative hit between f_1 and c_0 on a_0 (thick lines): $\sigma = \{f, c\}, \top = \{f_1 c_0\}, v = fc$. fc_{01} stands for the process corresponding to the state $f_0 c_1$ of the cooperating processes σ.

3.2   Stable State Pattern

Given a Process Hitting $(\Sigma, L, \mathcal{H})$, we prove the $|\Sigma|$-cliques of its Hitless Graph are exactly its stable states. Thus, stable states may be created by removing from the Process Hitting the very hits that make such patterns appear.

Theorem 1. *Let* $\mathcal{PH} = (\Sigma, L, \mathcal{H})$ *be a Process Hitting and* $\overline{\mathcal{PH}}$ *its Hitless Graph. A state* $s \in L$ *is stable if and only if s is a $|\Sigma|$-clique for* $\overline{\mathcal{PH}}$.

Proof. By definition, $next(s) = \emptyset$ if and only if there is no hit between any couple of processes present in s. This is equivalent to have s a clique of $\overline{\mathcal{PH}}$.

Figure 3 shows an instance of Process Hitting having one stable state.

We outline an algorithm for finding the n-cliques of a Hitless Graph $\overline{\mathcal{PH}} = (V, E)$ where $n = |\Sigma|$.

Thanks to Prop. 1, we split V into n partitions corresponding to each process: $V = \cup_{a \in \Sigma} V_a$, $V_a \subseteq L_a$. If one of this partition is empty, there can not be n-cliques as it requires to have at least one vertex in each partition. We will assume $V_a \neq \emptyset$, $\forall a \in \Sigma$.

For each partition $a \in \Sigma$ and each vertex $a_i \in V_a$, we define $E_{a_i}^b = \{b_j \in V_b \mid (a_i, b_j) \in E\}$ for each other partition $b \in \Sigma, b \neq a$, the set of vertices in V_b related to a_i. If there exists $b \in \Sigma$ such that $E_{a_i}^b = \emptyset$, the vertex a_i is removed from candidates as it can not belong to a n-clique. Finally, we set $E_{a_i}^a = \{a_i\}$.

Once this pruning is performed, we enumerate potential n-cliques. To reduce this enumeration, we choose the partition a sharing the least number of edges. For each vertex $a_i \in V_a$ we test for all $s \in \prod_{b \in \Sigma} E_{a_i}^b$ if s is a clique of $\overline{\mathcal{PH}}$.

For instance in Figure 3(b), a_1 is removed from the Hitless Graph ($E_{a_1}^b = \emptyset$), the partition associated to c is chosen (involved into only 4 edges), two states are tested: $a_0 b_0 c_1$ and $a_2 b_1 c_0$ and the latter reveals to be the only 3-clique.

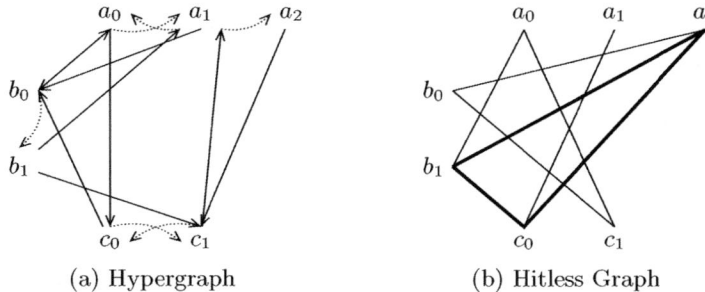

(a) Hypergraph   (b) Hitless Graph

Fig. 3. A Process Hitting represented by its Hypergraph (a) and its Hitless Graph (b). The Hitless Graph contains only one 3-clique between a_2, b_1 and c_0 (thick lines): this is the only stable state of this system.

4   From Process Hitting to René Thomas' Parameters

A René Thomas' discrete parameter gives the attractor levels for a gene when its regulators are in a given configuration. Many frameworks and tools dedicated to the study of GRNs take the full set of René Thomas' parameters as essential input [1,9,11]. In this section, we give a formal method to infer René Thomas' parameters for a GRN modelled in the Process Hitting framework.

Let (Γ, E_+, E_-) be a GRN graph. A René Thomas' parameter $K_{a,A,B}$, $a \in \Gamma, A \cup B \subseteq \Gamma$, $A \cap B = \emptyset$, gives the interval of attracting levels for a when genes in A are activating a and genes in B are inhibiting a. In this configuration, if the level of a is in $K_{a,A,B}$, then it will never change; otherwise the level of a will tend to a level in $K_{a,A,B}$.

Let $(\Sigma, L, \mathcal{H})$ be a Process Hitting where sorts are standing either for genes or for cooperative sorts, i.e. $\Sigma = \Gamma \cup \{\sigma^1, \ldots, \sigma^u\}$ with $\forall 1 \leq v \leq u$, $\sigma^v \subset \Gamma$. Let $K_{a,A,B}$ be the René Thomas' parameter to infer. For each sorts $b \in A \cup B$, its context $C^b_{a,A,B}$ is defined as the subset of processes L_b imposed by the René Thomas' parameter: if $b \in A$ (resp. B) only processes corresponding to positive (resp. negative) effective levels (Def. 2) are allowed. For each process $b \in \Gamma$ not regulating a (i.e. $b \notin A \cup B$), its context $C^b_{a,A,B}$ is simply L_b (7). The context $C^\sigma_{a,A,B}$ for cooperative sorts $\sigma \in \{\sigma^1, \ldots, \sigma^u\}$ is the set of states of its representatives in their context (8).

$$\forall b \in \Gamma, \ C^b_{a,A,B} = \begin{cases} levels_+(b, a) & \text{if } b \in A, \\ levels_-(b, a) & \text{if } b \in B, \\ L_b & \text{otherwise.} \end{cases} \quad (7)$$

$$\forall \sigma \in \{\sigma^1, \ldots, \sigma^u\}, \ C^\sigma_{a,A,B} = \{\sigma_\varsigma \mid \forall \varsigma \in \prod_{b \in \sigma} C^b_{a,A,B}\} \ . \quad (8)$$

We denote $\mathcal{H}_{a,A,B}$ the subset of the set of actions $\mathcal{H}$ on a that may be performed by processes of the context of any sort (9). A process of sort a is reachable if it belongs to the context of a or is the result of any action in $\mathcal{H}_{a,A,B}$. The set of such processes is noted $L^?_{a,A,B}$ (10). The set of reachable processes of sort a not hit by any other processes is noted $L^*_{a,A,B}$ (11). Thus, as long as present processes are in the context of their sort, if a process of sort a is in $L^*_{a,A,B}$, it will not be bounced. In this way, $L^*_{a,A,B}$ is called the set of *focal processes* of a.

$$\mathcal{H}_{a,A,B} = \{b_i \rightarrow a_j \upharpoonright a_k \in \mathcal{H} \mid b_i \in C^b_{a,A,B} \wedge a_j \in C^a_{a,A,B}\} \tag{9}$$

$$L^?_{a,A,B} = C^a_{a,A,B} \cup \{a_k \mid \forall b_i \rightarrow a_j \upharpoonright a_k \in \mathcal{H}_{a,A,B}\} \tag{10}$$

$$L^*_{a,A,B} = L^?_{a,A,B} \setminus \{a_j \mid \forall b_i \rightarrow a_j \upharpoonright a_k \in \mathcal{H}_{a,A,B}\} . \tag{11}$$

Finally, we check that the focal processes are attractors, i.e. all actions $\mathcal{H}_{a,A,B}$ make processes of sort a bounce in the direction of the focal processes. If such a condition is satisfied, the focal processes correspond to the value of the requested René Thomas' parameter. We point up that all these operations are linear with the number of actions in the Process Hitting.

Condition 1 (Focal processes are attractors)
$\forall b_i \rightarrow a_j \upharpoonright a_k \in \mathcal{H}_{a,A,B}, \ \forall a_f \in L^*_{a,A,B}, \ |f - k| < |f - j| .$

Property 3 If $L^*_{a,A,B}$ satisfies Cond. 1, $L^*_{a,A,B}$ is an interval.

Proof If $L^*_{a,A,B} = \{a_f, \ldots, a_{f'}\}$ is not an interval, there exists $b_i \rightarrow a_j \upharpoonright a_k \in \mathcal{H}_{a,A,B}$ such that $f < j < f'$. If Cond. 1 applies, we have $|f - k| < |f - j| \Rightarrow k < j \Rightarrow |f' - k| > |f' - j|$ which contradicts Cond. 1.

Theorem 2. *If* $L^*_{a,A,B} \neq \emptyset$ *and Cond. 1 holds, then* $K_{a,A,B} = L^*_{a,A,B}$.

Proof. By construction of $L^*_{a,A,B}$ and application of Cond. 1 and Prop. 3, it immediately appears that if $L^*_{a,A,B} \neq \emptyset$, it is the set of attracting levels for a.

Consequently, there might exist configurations without any correspondence with René Thomas' parameters. First, $L^*_{a,A,B} = \emptyset$ means the gene a is unstable in the fixed context, i.e. its level is changing forever. Second, Cond. 1 is violated when there exists opposite focal processes, i.e. the fate of a is not deterministic.

One of the main reasons for non-determinism of Process Hitting is the absence of cooperativity between hits to a same target which may then independently be bounced to both higher and lower processes. We leave as an open question the problem to know whether such unstable and/or non-deterministic dynamics are biologically relevant.

5   Temporal and Stochastic Parameters

Further dynamics refinements may be achieved by taking into account the temporal and stochastic dimensions of biological reactions. On the one hand, we may consider the probability of a reaction to occur at a given state. By introducing stochastic parameters into discrete models, we aim at computing the probability of observing an expected behaviour. On the other hand, because they are faster, some reactions always apply before others. By introducing temporal parameters into discrete models, we aim at reducing their dynamics to match such behaviours.

5.1   From Process Hitting to the Stochastic π-Calculus

The Stochastic π-Calculus [20] adds the capability to attach *use rates* to channels and internal actions of the π-Calculus. This gives a natural introduction for temporal and stochastic aspects in our Process Hitting framework.

A use rate controls both the duration and the probability of a reaction (communication on a channel or internal action). It is associated to a probability distribution for firing reaction along the time. The usual probability distribution is the exponential one, allowing efficient simulations through a Gillespie-like algorithm [12,21]. This is the one we consider for the rest of this paper.

The probability along time t of firing a reaction with use rate r is given by $F(t) = 1 - e^{-rt}$. The average duration of this reaction is r^{-1} with a variance of r^{-2}. When x reactions are possible having use rates of $r_1, \ldots, r_x$ respectively, the probability that the yth reaction is fired is given by $\frac{r_y}{r_1 + \cdots + r_x}$.

The translation of Process Hitting $(\Sigma, L, \mathcal{H})$ into the Stochastic π-Calculus is the same as the one presented in Section 2.4. Additionally, to each channel γ_α, or internal action τ_α, a use rate r_α is attached.

5.2   Stochasticity Absorption

Use rates are both temporal and stochastic parameters. Nonetheless, these two aspects are closely tied: the lower a use rate is, the higher the variance around its mean duration is. We introduce a *stochasticity absorption factor* to control this variance to favour either the stochastic or the temporal behaviour of an action.

We propose to replace the exponential distribution of a reaction with a rate r by the distribution of the sum of sa random variables each having an exponential distribution of parameter $r.sa$. The resulting probability distribution is also known as the *Erlang distribution*. The average duration is unchanged: $(r.sa)^{-1}sa = r^{-1}$, but the variance is divided by sa: $(r.sa)^{-2}sa = r^{-2}sa^{-1}$. sa stands for the stochasticity absorption factor. Based on the previously presented translation from the Process Hitting to the Stochastic π-Calculus, we supply a simple method to achieve this stochasticity absorption factor which do not require to adapt simulation algorithms based on the memoryless property of the exponential law [22].

Basically, to each channel γ_α, or internal action τ_α, a use rate r_α and a stochasticity absorption factor sa_α is attached. To each component α of the sum defined by the π-Calculus process A_i in the expressions (2),(3), a counter c_α is attached, initially, $c_\alpha = 1$. This counter is given as a parameter for A_i. As long as this counter is not equal to sa_α, A_i is restarted and the counter is incremented by one. When the counter reaches the stochasticity absorption factor value, the next process replaces A_i, having all its counters reset to 1. Let $(\Sigma, L, \mathcal{H})$ be a Process Hitting, for each process a_i of $\mathcal{PH}$, a π-Calculus process A_i is defined as follows.

$$A_i(\tilde{c}) ::= \sum_{\substack{\alpha = a_i \to b_j \,\uparrow\, b_k \in \mathcal{H} \\ a \neq b}} !\gamma_\alpha . A_i(\tilde{c})$$

$$+ \sum_{\substack{\alpha = b_j \to a_i \,\uparrow\, a_k \in \mathcal{H} \\ a \neq b}} [c_\alpha < sa_\alpha]?\gamma_\alpha . A_i(\tilde{c}[c_\alpha + 1]) + [c_\alpha = sa_\alpha]?\gamma_\alpha . A_k(\tilde{1})$$

$$+ \sum_{\alpha = a_i \to a_i \,\uparrow\, a_k \in \mathcal{H}} [c_\alpha < sa_\alpha]\tau_\alpha . A_i(\tilde{c}[c_\alpha + 1]) + [c_\alpha = sa_\alpha]\tau_\alpha . A_k(\tilde{1})$$

where $\hat{c} = c_1, \ldots, c_n$ with $n = |\{b_j \to a_i \,\uparrow\, a_k \in \mathcal{H}\}|$. $\hat{c}[c_\alpha + 1] = c_1, \ldots, c_\alpha + 1, \ldots, c_n$. $A_k(\tilde{1})$ is an abbreviation for the recursive call to A_k with all parameters set to 1. $[cond]\pi.P$ stands for an action π enabled only when $cond$ is satisfied.

6   Applications

6.1   Metazoan Segmentation

In this section, we illustrate our method and its benefits on a case study in which our aim is to control the final state of the corresponding GRN. The GRN we chose has been established *in silico* by François et al. [23] but in a differential equations framework. It aims at generalizing a common motif present in biological segmentation networks such as the *Drosophila*.

The GRN (Figure 4(a)) is composed of three genes. A wavefront gene f activates the gap-gene a whose products are responsible or stripes. Gene f also activates a gene c whose products repress the gene a. The auto-inhibition of c generalizes a chain of repressors on a. The apparition of stripes has to be regular. We attach to each gene two processes representing their qualitative levels (missing or present) — for instance c_0 (absence) and c_1 (presence) are processes for c. When f switches off, c goes to process c_0 and a has two fates, ending either at process a_0 or a_1. We are interested in controlling the final process for a.

The Process Hitting for generalised dynamics of the GRN (Section 2) is computed first. Figure 4(b) shows its hypergraph. The specification of dynamics implies two cooperative hits in the Process Hitting: first, c_0 needs products of f to bounce to process c_1; second, expression of a only increases if both f activates it (i.e. process f_1 is present) and c does not inhibit it (i.e. c_0 is present). Consequently, we create a cooperative sort fc reflecting the state of f and c (Section 3.1) and replace the independent hits from c_0 and f_1 to c_0 and a_0 by hits from fc_{10}. The resulting Process Hitting is represented in Figure 5(a).

By looking at the Hitless Graph of the Process Hitting (Figure 5(b)), only one stable state is present: $f_0 c_0 fc_{00} a_0$. The stability of the state $f_0 c_0 fc_{00} a_1$ is controlled by the absence of inhibition by f_0 on a_1. By removing the action

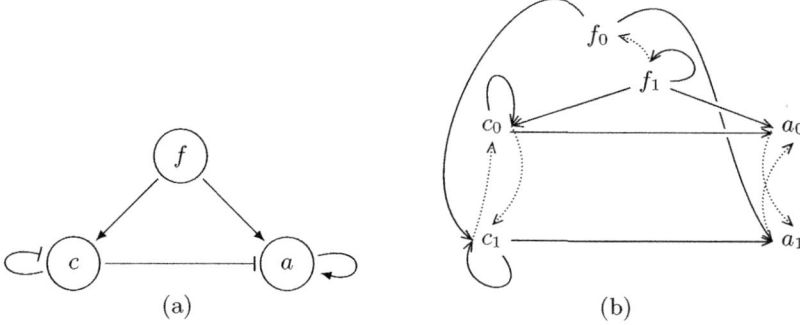

(a)   (b)

Fig. 4. The starting Gene Regulatory Network graph (a), arrow-ended edges represent the positive regulations, and bar-ended edges the negative ones. All regulation thresholds are 1. The Process Hitting (b) for its generalized dynamics. Cooperativity between f_1 and c_0 on a_0 and c_0 will be applied in the same way as in example. 1.

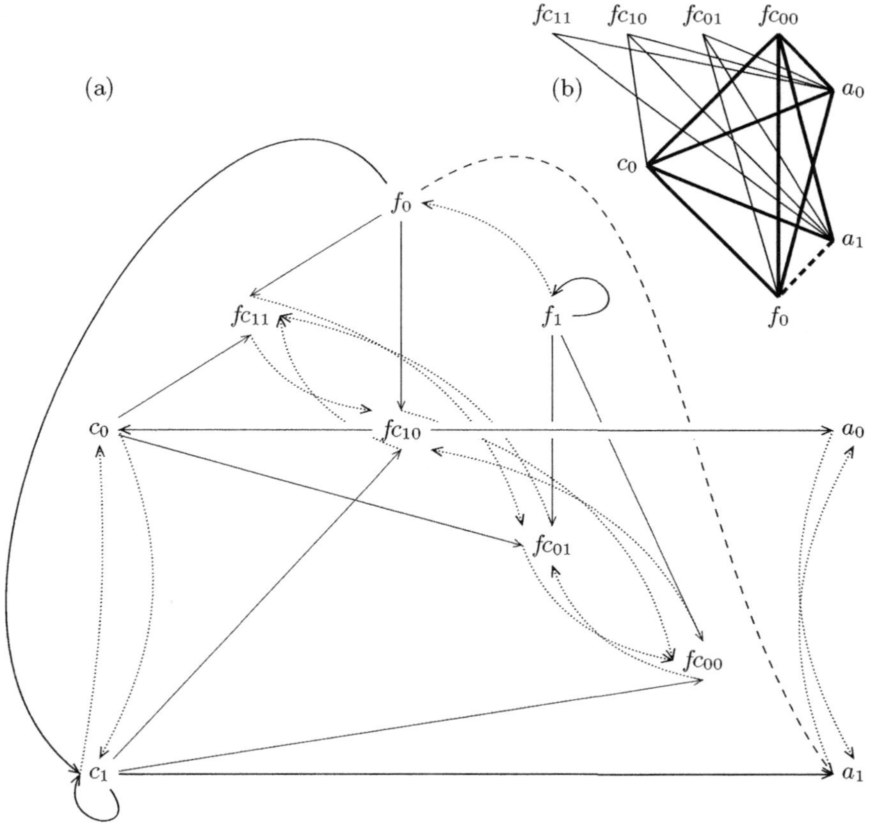

Fig. 5. The final Process Hitting (a) resulting from the refinement of the generalized dynamics depicted on Fig. 4(b). Cooperativity between f_1 and c_0 on a_0 and c_0 has been built in the same way as in example 1. Absence of hit from f_0 to a_1 (dashed lines) controls the presence of the relation between f_0 and a_1 in the Hitless Graph (b). If such a relation exists, two 4-cliques are presents: $c_0 f_0 f c_{00} a_0$ and $c_0 f_0 f c_{00} a_1$ (thick lines).

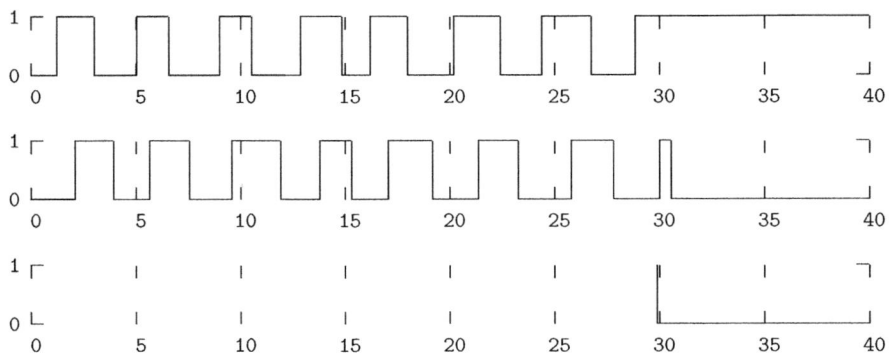

Fig. 6. Simulation of the Process Hitting for segmentation: evolution of the expressions of the gap-gene a (top), the autonomous clock c (middle) and the wavefront f

$f_0 \rightarrow a_1 \, \Gamma \, a_0$ from the Process Hitting, we make the state $f_0 c_0 f c_{00} a_1$ stable. The full set of corresponding René Thomas' parameters for the genes a and c is inferred by applying the method depicted in Section 4. We get:

$$
\begin{array}{lll}
K_{a,\emptyset,\{a,c,f\}} = 0 & K_{a,\{a,c\},\{f\}} = 1 & K_{c,\emptyset,\{c,f\}} = 0 \\
K_{a,\{a\},\{c,f\}} = 0 & K_{a,\{a,f\},\{c\}} = 0 & K_{c,\{c\},\{f\}} = 0 \\
K_{a,\{c\},\{a,f\}} = 0 & K_{a,\{c,f\},\{a\}} = 1 & K_{c,\{f\},\{c\}} = 0 \\
K_{a,\{f\},\{a,c\}} = 0 & K_{a,\{a,c,f\},\emptyset} = 1 & K_{c,\{c,f\},\emptyset} = 1 \ .
\end{array}
$$

We are interested in controlling the final process of sort a — either a_0 or a_1 — when f goes down to f_0. Looking at the Process Hitting hypergraph on Figure 5(a) and considering f_0 is present, we deduce that the more c_1 is present, the more a_1 may be hit by c_1 to bounce to a_0; similarly, the more fc_{10} is present, the more a_0 may be hit by it to bounce to a_1. We tune actions only triggered by f_0: we reduce the presence of c_1 by increasing the rate of the action $f_0 \rightarrow c_1 \, \Gamma \, c_0$ and extend the presence of fc_{10} by reducing the rate of $f_0 \rightarrow fc_{10} \, \Gamma \, fc_{00}$. This leads to an increase of the probability for a to bounce to process a_1.

Finally, to obtain regular stripes, we set a high stochasticity absorption factor to actions responsible of the bounces of processes of sort c and a when f_1 is present. Figure 6 plots the evolution of the genes a,c and f during a simulation under SPiM [24] of the Process Hitting illustrated by Figure 5(a) with initial state $f_1 c_0 fc_{10} a_0$ and a fast rate for the action $f_0 \rightarrow c_1 \, \Gamma \, c_0$ compared to the rate of $c_1 \rightarrow a_1 \, \Gamma \, a_0$. The rate values have been arbitrarily choosen and respect the infered relations between them. Appendix B.1 details the Process Hitting used for the simulation.

From the obtained simulation trace, we observe that f_0 hits c_1 before c_1 had time to hit a_1: the final state is then $f_0 c_0 fc_{00} a_1$.

Thanks to the Process Hitting framework, it has been easy to build a qualitative model of the biological system by refining the generalised dynamics of the GRN. Using a simple reasoning on the Process Hitting structure, relation between regulation delays to favour a final state have been infered. These results are coherent with those obtained using differential equations as done in [23].

6.2 ERBB Receptor-Regulated G1/S Transition

The aim of this section is to demonstrate the scalability of the refining approach on Process Hittings modelling large GRNs.

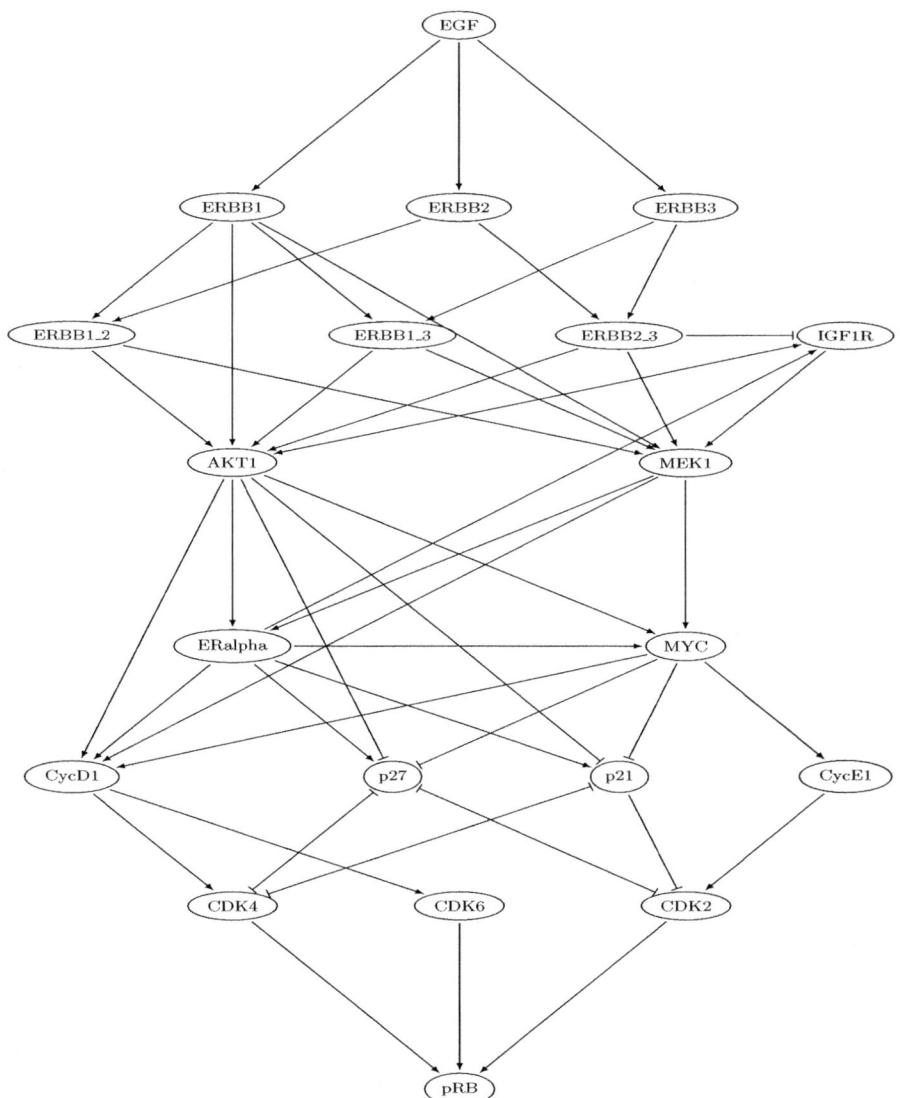

Fig. 7. ERBB receptor-regulated G1/S transition GRN reproduced from [25, Figure 3]

The selected GRN relates regulations between 20 genes. This GRN models the *ERBB receptor-regulated G1/S transition* involved in the breast cancer. It has been extracted from published data by Sahin et al. [25] and is reproduced in Figure 7. This network acts as an activation cascade for the gene *pRB* which controls the G1/S transition involved in cell divisions. The gene *EGF* then acts as an input of this cascade: when expressed, *pRB* will be potentially activated. Based on the literature, Sahin et al. have also established a set of logical rules controlling the activation of the various genes present in the network.

Starting from the GRN, its generalised dynamics expressed in Process Hitting is first computed. Then, cooperations between the different sorts are built from the logical rules. The Process Hitting obtained contains 670 actions and stands for 2^{64} ($\approx 2.10^{19}$) states that has hopefully not be built. This model is fully detailled in Appendix B.2.

The computation of all the stable states present in the dynamics is done using the algorithm sketched by Section 3.2. It results in 5 stables states (also detailed in the appendix) that are computed in less than one second.

It is worth noticing that no assumption is made on the initial state of the system. *All* the stable states of the model are computed. This is a major difference with the approach presented in [25] where only dynamics starting from a fixed state can be studied.

7   Conclusion

We introduced the Process Hitting framework for modelling qualitative dynamics of GRNs with temporal and stochastic features. Temporal and stochastic parameters determine probabilities, durations and temporal variance of reactions in the model. We exhibited a direct translation from Process Hitting to the Stochastic π-Calculus. Detection of stable states and inference of René Thomas' parameters for dynamics derive from this framework. The methods we offered work by successive refinements of generalised dynamics for GRNs, by specifying both the cooperativity between genes and the expected stable states. We illustrated this method by inferring temporal parameters for the dynamics of a GRN generalizing metazoan segmentation processes (with the aim of controlling its final state). The scalability of the presented approach has been experimented on a Process Hitting modelling a GRN composed of 20 genes and computing its stable states.

The Process Hitting brings a formal framework for progressively adding knowledge of the dynamics of a GRN by refining an abstracted behaviour. The compositionality of the framework and the presence of particular structure patterns lead to scalable methods for dynamics analyses (stable states, René Thomas' parameters, etc.). Mainly, thanks to these Process Hitting patterns, it becomes possible to perform a local analysis, which has the major advantage to prevent us from exploring the full state and parameter space.

In future works, we aim at identifying more Process Hitting patterns leading to the emergence of particular behaviours (e.g. oscillations) and especially hybrid patterns coupling both discrete structure and continuous temporal and stochastic parameters. The verification of Process Hittings could be performed by using a translation into Petri Nets or into PRISM. Translating Process Hittings into more sophisticated process aglebras (Beta Workbench, Bio-PEPA, etc.) may also be of interest. Finally, techniques have to be developed to infer automatically temporal and stochastic parameters of Process Hittings modelling GRNs.

Supplementary Material

The Process Hitting compiler to SPiM, a stable states computer and presented models are available at the following URL:
http://www.irccyn.ec-nantes.fr/~pauleve/processhitting-refining.tar.gz

References

1. Richard, A., Comet, J.P., Bernot, G.: Formal Methods for Modeling Biological Regulatory Networks. In: Modern Formal Methods and Applications, pp. 83–122 (2006)
2. Danos, V., Feret, J., Fontana, W., Harmer, R., Krivine, J.: Rule-Based Modelling of Cellular Signalling. In: Caires, L., Li, L. (eds.) CONCUR 2007. LNCS, vol. 4703, pp. 17–41. Springer, Heidelberg (2007)
3. Priami, C., Regev, A., Shapiro, E., Silverman, W.: Application of a stochastic name-passing calculus to representation and simulation of molecular processes. Inf. Process. Lett. 80(1), 25–31 (2001)
4. Kuttler, C., Niehren, J.: Gene regulation in the pi calculus: Simulating cooperativity at the lambda switch. In: Priami, C., Ingólfsdóttir, A., Mishra, B., Riis Nielson, H. (eds.) Transactions on Computational Systems Biology VII. LNCS (LNBI), vol. 4230, pp. 24–55. Springer, Heidelberg (2006)
5. Blossey, R., Cardelli, L., Phillips, A.: A compositional approach to the stochastic dynamics of gene networks. In: Priami, C., Cardelli, L., Emmott, S. (eds.) Transactions on Computational Systems Biology IV. LNCS (LNBI), vol. 3939, pp. 99–122. Springer, Heidelberg (2006)
6. Popova-Zeugmann, L., Heiner, M., Koch, I.: Time petri nets for modelling and analysis of biochemical networks. Fundamenta Informaticae 67(1), 149–162 (2005)
7. Heiner, M., Gilbert, D., Donaldson, R.: Petri Nets for Systems and Synthetic Biology. In: Formal Methods for Computational Systems Biology, pp. 215–264 (2008)
8. Rizk, A., Batt, G., Fages, F., Soliman, S.: On a Continuous Degree of Satisfaction of Temporal Logic Formulae with Applications to Systems Biology. In: Computational Methods in Systems Biology, pp. 251–268 (2008)
9. Siebert, H., Bockmayr, A.: Incorporating Time Delays into the Logical Analysis of Gene Regulatory Networks. In: Computational Methods in Systems Biology, pp. 169–183 (2006)
10. Alur, R., Belta, C., Kumar, V., Mintz, M., Pappas, G.J., Rubin, H., Schug, J.: Modeling and analyzing biomolecular networks. Computing in Science and Engineering 4(1), 20–31 (2002)
11. Ahmad, J., Bernot, G., Comet, J.P., Lime, D., Roux, O.: Hybrid modelling and dynamical analysis of gene regulatory networks with delays. Complexus 3(4), 231–251 (2006)
12. Phillips, A., Cardelli, L.: Efficient, correct simulation of biological processes in the stochastic pi-calculus. In: Calder, M., Gilmore, S. (eds.) CMSB 2007. LNCS (LNBI), vol. 4695, pp. 184–199. Springer, Heidelberg (2007)

13. Dematte, L., Priami, C., Romanel, A.: The Beta Workbench: a computational tool to study the dynamics of biological systems. Brief Bioinform., bbn023 (2008)
14. Ciocchetta, F., Hillston, J.: Bio-pepa: A framework for the modelling and analysis of biological systems. Theoretical Computer Science 410(33-34), 3065–3084 (2009)
15. Hinton, A., Kwiatkowska, M., Norman, G., Parker, D.: PRISM: A tool for automatic verification of probabilistic systems. In: Hermanns, H. (ed.) TACAS 2006. LNCS, vol. 3920, pp. 441–444. Springer, Heidelberg (2006)
16. Norman, G., Palamidessi, C., Parker, D., Wu, P.: Model checking probabilistic and stochastic extensions of the π-calculus. IEEE Transactions on Software Engineering 35(2), 209–223 (2009)
17. Bernot, G., Cassez, F., Comet, J.P., Delaplace, F., Müller, C., Roux, O.: Semantics of biological regulatory networks. Electronic Notes in Theoretical Computer Science 180(3), 3–14 (2007)
18. Milner, R.: A calculus of mobile processes, parts. I and II. Information and Computation 100, 1–77 (1992)
19. Bernot, G., Comet, J.P., Khalis, Z.: Gene regulatory networks with multiplexes. In: European Simulation and Modelling Conference Proceedings, pp. 423–432 (October 2008)
20. Priami, C.: Stochastic π-Calculus. The Computer Journal 38(7), 578–589 (1995)
21. Gillespie, D.T.: Exact stochastic simulation of coupled chemical reactions. The Journal of Physical Chemistry 81(25), 2340–2361 (1977)
22. Priami, C.: Stochastic π-calculus with general distributions. In: Proc. of the 4th Workshop on Process Algebras and Performance Modelling, CLUT, pp. 41–57 (1996)
23. Francois, P., Hakim, V., Siggia, E.D.: Deriving structure from evolution: metazoan segmentation. Mol. Syst. Biol. 3 (2007)
24. Phillips, A.: SPiM, http://research.microsoft.com/~aphillip/spim
25. Sahin, O., Frohlich, H., Lobke, C., Korf, U., Burmester, S., Majety, M., Mattern, J., Schupp, I., Chaouiya, C., Thieffry, D., Poustka, A., Wiemann, S., Beissbarth, T., Arlt, D.: Modeling ERBB receptor-regulated G1/S transition to find novel targets for de novo trastuzumab resistance. BMC Systems Biology 3(1) (2009)

A   Process Hitting Related Tools

This appendix briefly presents currently available implementations of tools manipulating Process Hittings. They are available at the following URL:

http://www.irccyn.ec-nantes.fr/~pauleve/processhitting-refining.tar.gz

Implemented in the OCAML language, these tools have a command-line user interline.

A.1   Process Hitting Specification

A basic language has been setup to specify Process Hitting using a text file. Main features are presented here, more details can be found in the provided archive. Full examples of Process Hitting specifications are given in Appendix B.

Sort definition. A sort is declared by giving the process with the highest rank (i.e. a_{l_a} for the sort a, with the notations used in Section 2).

```
process a X
```

Action specification. An action $a_i \rightarrow b_j \upharpoonleft b_k$ is added by the following instruction:

```
a i -> b j k @ rate ~ absorption
```

Generalised dynamics of GRNs. The GRN macro computes the generalised dynamics of the specified GRN according to Section 2.5. An activating (resp. inhibiting) edge $a \xrightarrow{X} b$ from gene a to gene b active with a threshold X is noted as a x -> + b (resp. a x -> - b).

Hereafter is an instance of the use of the GRN macro for the GRN having activating edges $a \xrightarrow{1} b$ and $a \xrightarrow{2} a$, and inhibiting edge $b \xrightarrow{1} b$.

```
GRN([a 1 -> + b; b 1 -> - a; a 2 -> + a ])
```

Refinement: cooperations. The COOPERATIVITY([a1;...;an] -> b j k, [s1;...;sp]) macro creates a cooperative sort $\sigma = \{a^1, \ldots, a^n\}$ for the bounce $b_j \upharpoonleft b_k$. This cooperation is effective for every state $s_1, \ldots, s_p$.

The following instruction creates a cooperativity between sorts a and b to bounce process c_0 to c_1 only if $a_1 b_0$ or $a_0 b_1$ are present.

```
COOPERATIVITY([a;b] -> c 0 1, [[1;0];[0;1]])
```

Refinement: action removing An action $a_i \rightarrow b_j \upharpoonleft b_k$ can be deleted by using the macro RM:

```
RM({a i -> b j k})
```

A.2   Compiler to SPiM

This tool translates a Process Hitting specification into a SPiM model according to Section 5.2.

```
phc -spim <model.ph> <output.spi>
```

A.3   Stable States Listing

The stable states of a Process Hitting are determined used an implementation of the algorithm sketched in Section 3.2. This algorithm computes the n-cliques of the hitless graph for the given Process Hitting, where n is the number of sorts. The efficiency of this computation heavily relies on the order of sorts when building cliques. Currently, basic heuristics are used to select the order of the sorts. More sophisticated analyses may conduct to dramatically improve the efficiency of this algorithm.

This tool is compiled into the executable ph-stable-states and takes as argument the filename of the Process Hitting specification :

```
ph-stable-states <model.ph>
```

B   Process Hitting Examples

This appendix details the Process Hittings used in Section 6. They are specified in the language presented in the previous appendix.

B.1   Metazoan Segmentation

The following Process Hitting models the metazoan segmentation presented in Section. 6.1. Actions are specified separately and rates have been assigned to values matching the relations infered in the application case study. The directive sample and initial_state instructions are of use for SPiM only. A result of the execution of this Process Hitting translated into SPiM is given by Figure 6.

```
directive sample 40.

process a 1 process c 1 process f 1
process fc 3 (* cooperative sort {f,c} *)
c 1 -> fc 0 1 @5.
c 1 -> fc 2 3 @5.
c 0 -> fc 1 0 @10.
c 0 -> fc 3 2 @5.
f 1 -> fc 0 2 @10.
f 1 -> fc 1 3 @10.
f 0 -> fc 2 0 @0.1
f 0 -> fc 3 1 @0.1

(* actions on c *)
fc 2 -> c 0 1 @0.5~50 (* only if (f1,c0) *)
c 1 -> c 1 0 @0.5~50
(* actions on a *)
fc 2 -> a 0 1 @1.~50 (* only if (f1,c0) *)
c 1 -> a 1 0 @1.~50
(* actions on f *)
f 1 -> f 1 0 @0.034~100 (* auto-off *)
f 0 -> c 1 0 @0.1

initial_state f 1, c 0, a 0
```

B.2   ERBB Receptor-Regulated G1/S Transition

The following Process Hitting results from the case study presented in Section 6.2. It starts by specifying the GRN depicted by Figure 7 to compute its generalised dynamics. The logical rules setup by Sahin et al. [25] are then applied by using sorts cooperativity.

This Process Hitting contains 670 actions and 2^{64} ($\approx 2.10^{19}$) states. Only 5 stable states exist and are determined in less than a second using the tool presented in the previous appendix.

Below is the list of stable states present in the Process Hitting. For each stable state, only genes at level 1 are written.

— AKT1, CDK2, CDK4, CDK6, CycD1, CycE1, EGF, ERBB1, ERBB1_2, ERBB1_3, ERBB2, ERBB2_3, ERBB3, ERalpha, MEK1, MYC, pRB.
— AKT1, CDK2, CDK4, CDK6, CycD1, CycE1, ERalpha, IGF1R, MEK1, MYC, pRB.
— AKT1, CDK2, CycE1, EGF, ERBB1, ERBB1_2, ERBB1_3, ERBB2, ERBB2_3, ERBB3, ERalpha, MEK1, MYC.
— AKT1, CDK2, CycE1, ERalpha, IGF1R, MEK1, MYC.
— ∅ (all genes have level 0).

```
process AKT1 1
process CDK2 1 process CDK4 1 process CDK6 1
process CycD1 1 process CycE1 1
process EGF 1 process ERalpha 1
process ERBB1 1 process ERBB1_2 1 process ERBB1_3 1
process ERBB2 1 process ERBB2_3 1 process ERBB3 1
process IGF1R 1 process MEK1 1 process MYC 1
process p21 1 process p27 1
process pRB 1

GRN([
    ERBB2_3 1 -> + AKT1; ERBB2_3 1 -> + MEK1; ERBB2_3 1 -> - IGF1R;
    ERBB2 1 -> + ERBB2_3; ERBB2 1 -> + ERBB1_2; ERBB3 1 -> + ERBB2_3;
    ERBB3 1 -> + ERBB1_3;
    CycE1 1 -> + CDK2;
    MEK1 1 -> + CycD1; MEK1 1 -> + ERalpha; MEK1 1 -> + MYC;
    CDK4 1 -> + pRB; CDK4 1 -> - p21; CDK4 1 -> - p27;
    ERalpha 1 -> + CycD1; ERalpha 1 -> + IGF1R; ERalpha 1 -> + p21;
    ERalpha 1 -> + MYC; ERalpha 1 -> + p27;
    MYC 1 -> + CycE1; MYC 1 -> - p21; MYC 1 -> + CycD1; MYC 1 -> - p27;
    CDK6 1 -> + pRB;
    ERBB1 1 -> + ERBB1_2; ERBB1 1 -> + ERBB1_3; ERBB1 1 -> + AKT1;
    ERBB1 1 -> + MEK1;
    IGF1R 1 -> + AKT1; IGF1R 1 -> + MEK1;
    ERBB1_3 1 -> + AKT1; ERBB1_3 1 -> + MEK1;
    p27 1 -> - CDK2; p27 1 -> - CDK4;
    CDK2 1 -> - p27; CDK2 1 -> + pRB;
    p21 1 -> - CDK2; p21 1 -> - CDK4;
    CycD1 1 -> + CDK4; CycD1 1 -> + CDK6;
    EGF 1 -> + ERBB1; EGF 1 -> + ERBB2; EGF 1 -> + ERBB3;
    AKT1 1 -> + CycD1; AKT1 1 -> + MYC; AKT1 1 -> - p27; AKT1 1 -> + ERalpha;
    AKT1 1 -> + IGF1R; AKT1 1 -> - p21;
    ERBB1_2 1 -> + AKT1; ERBB1_2 1 -> + MEK1;
])

COOPERATIVITY([ERBB1;ERBB2] -> ERBB1_2 0 1, [[1;1]])
COOPERATIVITY([ERBB1;ERBB3] -> ERBB1_3 0 1, [[1;1]])
COOPERATIVITY([ERBB2;ERBB3] -> ERBB2_3 0 1, [[1;1]])

COOPERATIVITY([ERBB2_3;AKT1] -> IGF1R 0 1, [[0;1]])
COOPERATIVITY([ERBB2_3;ERalpha] -> IGF1R 0 1, [[0;1]])
COOPERATIVITY([AKT1;ERalpha] -> IGF1R 1 0, [[0;0]])

COOPERATIVITY([AKT1;MEK1] -> ERalpha 1 0, [[0;0]])
COOPERATIVITY([AKT1;MEK1;ERalpha] -> MYC 1 0, [[0;0;0]])
COOPERATIVITY([ERBB1;ERBB1_2;ERBB1_3;ERBB2_3;IGF1R] -> AKT1 1 0,
    [[0;0;0;0;0]])
COOPERATIVITY([ERBB1;ERBB1_2;ERBB1_3;ERBB2_3;IGF1R] -> MEK1 1 0,
    [[0;0;0;0;0]])
COOPERATIVITY([CycE1;p21;p27] -> CDK2 0 1, [[1;0;0]])
COOPERATIVITY([CycD1;p21;p27] -> CDK4 0 1, [[1;0;0]])
COOPERATIVITY([ERalpha;MYC;AKT1;MEK1] -> CycD1 0 1, [[1;1;1;0];[1;1;0;1]])
COOPERATIVITY([AKT1;MEK1] -> CycD1 1 0, [[0;0]])
COOPERATIVITY([ERalpha;AKT1;MYC;CDK4] -> p21 0 1, [[1;0;0;0]])
COOPERATIVITY([ERalpha;CDK4;CDK2;AKT1;MYC] -> p27 0 1, [[1;0;0;0;0]])
COOPERATIVITY([CDK2;CDK4;CDK6] -> pRB 0 1, [[0;1;1];[1;1;1]])
RM({CDK2 0 -> pRB 1 0})
RM({EGF 1 -> EGF 1 0}) (* prevent self-degradation (input) *)
```

Author Index